From Byron to bin Laden

From Byron to bin Laden

A History of Foreign War Volunteers

NIR ARIELLI

Harvard University Press

Cambridge, Massachusetts
London, England
2017

Library of Congress Cataloging-in-Publication Data
Names: Arielli, Nir, 1975– author.
Title: From Byron to bin Laden : a history of foreign war volunteers / Nir Arielli.
Description: Cambridge, Massachusetts : Harvard University Press, 2018. |
Includes bibliographical references and index.
Identifiers: LCCN 2017017573 | ISBN 9780674979567 (alk. paper)
Subjects: LCSH: Foreign enlistment—History. | Military service, Voluntary—History. |
Soldiers—Psychology.
Classification: LCC UB321 .A75 2018 | DDC 355.2/2362—dc23
LC record available at https://lccn.loc.gov/2017017573

Contents

From Byron to bin Laden

Introduction

In August 2014 a masked man, dressed in black and holding a knife, was filmed delivering a propaganda message and then executing James Foley, a reporter from the United States who was being held hostage by the Islamic State in Syria. At first the identity of the executioner remained a mystery, although his accent betrayed that he had been raised in London. Soon the tabloid press in Britain reported that the man was part of a "gang of British jihadis known as 'The Beatles,'" therefore bestowing on him the nickname "Jihadi John."[1] It was only in February 2015 that the masked man, who had meanwhile been filmed executing other foreign hostages, was formally identified as Mohammed Emwazi.

A native of Kuwait and a British citizen, Emwazi grew up in west London and studied computer programing at the University of Westminster.[2] He reportedly came to the attention of the British Secret Service MI5 and other security agencies in 2009 because of his association with Islamists who were linked with the organization Al-Shabaab in Somalia. His relatives would later claim that Emwazi's "encounters with the British security services had wrecked his life in Britain and had ended his plans for marriage and work in his homeland of Kuwait."[3] The Islamic State, on the other hand, presented Emwazi as a devout and committed man, endowed with "foresight and decisiveness," and not as a person rattled by the surveillance he had been placed under. In the organization, Emwazi was given the *kunya* or nom de guerre Abu Muharib al-Muhajir (the name Muharib means "warrior" and Muhajir, in this context, denotes a jihadi immigrant). The Islamic State's English-language magazine, *Dabiq,* took pride in the fact that Emwazi had slipped out of the United Kingdom and reached Syria undetected, "right under the nose of the much-overrated MI5."[4] In January 2016 the organization confirmed that Emwazi had been killed by a drone strike on 12 November 2015.

Another British man who made his way to Syria to fight in its civil war is Michael Enright, a British actor who had a minor role in the Hollywood film *Pirates of the Caribbean: Dead Man's Chest.* But unlike Emwazi, Enright joined the People's Protection Unit or YPG, a Kurdish military organization that currently fights against the Islamic State. In fact, Enright alluded to Emwazi as one of the reasons for his decision to volunteer in an interview he gave Al-Aan TV in early June 2015: "I particularly became aware of it [the Islamic State] when they cut off an American journalist's head . . . and what was even worse for me was that it was an Englishman who did it. . . . I've got to try to right that wrong."[5]

While both men made international headlines, it is Islamic foreign fighters like Emwazi who have captivated the attention of politicians, security agencies, and public opinion more broadly. In contemporary debates, foreign fighters are frequently seen as a problem. World leaders from different countries, who often disagree with each other on many issues—including how to respond to the civil war in Syria—share a concern about the dangers posed by returning, radicalized veterans of the Syrian conflict. For instance, former British prime minister Tony Blair declared in June 2014 that "already the security agencies of Europe believe our biggest future threat will come from returning fighters from Syria."[6] Russian president Vladimir Putin expressed similar concerns in an article published in the *New York Times* in September 2013. The presence of foreign fighters in Syria, he wrote, which includes "hundreds of militants from Western countries and even Russia, are an issue of our deep concern. Might they not return to our countries with experience acquired in Syria? . . . This threatens us all."[7] Russia's foreign minister, Sergei Lavrov, later echoed Putin, warning of "the militants" from Russia and other former Soviet states, "whose number amounts to thousands" and who "will return and will do their black deeds in our country."[8]

Zhang Chunxian, the Communist Party secretary of Xinjiang in China, likewise warned of the danger posed by those Uyghurs (a Muslim ethnic group from western China) who "have fled overseas and joined Isis." This, he said, has a direct bearing on China because "some who fought [in Syria] returned to Xinjiang to participate in terrorist plots."[9] Meanwhile, US president Barack Obama also cautioned that "foreign fighters were likely to return to their home countries to carry out attacks."[10] With such an international consensus, it is unsurprising that the United Nations (UN) Security Council unanimously adopted Resolution 2178 in September 2014. In this resolution the Security Council expressed concern that what it called "foreign terrorist fighters" may "pose a serious threat to their States of origin, the States they transit and the States to which they travel."[11]

The aim of this book is to examine foreign-war volunteers not merely as a "problem." It makes a historian's contribution to contemporary debates by tracing the roots of the phenomenon to the late eighteenth century and identifying its main attributes in subsequent generations. Among other things, the book shows how foreign volunteers were perceived differently by different people in different historical contexts. While today, as the statements above illustrate, there is a tendency to conflate foreign fighters with terrorists, people who traveled abroad to voluntarily fight in foreign conflicts have also been hailed as heroes. As we will see, this was the case with the Marquis de Lafayette, a Frenchman who fought in the American Revolution in the 1770s, and the British poet Lord Byron, who took part in the Greek War of Independence in the 1820s. When discussing the topic of my research, many of my interlocutors tended to express positive views about certain groups of foreign volunteers such as the International Brigades that fought in the Spanish Civil War (1936–1939). How one perceives the cause foreign volunteers fought for largely determines one's positive or negative appraisal of such individuals. In this book, however, I have tried, as far as possible, to avoid passing judgment on "good" or "bad" and "just" or "unjust" causes.

The study of the history of foreign volunteering opens up a broad range of questions that relate to human motivations, ideology, gender, state–citizen relations, international law, military effectiveness, radicalization, the memory of war, and many more. To address these questions, I have made use of evidence from both primary and secondary sources pertaining to the participation of foreigners in different kinds of armed conflicts that span almost two and a half centuries. There are vast differences between the cases under examination. For instance, foreign volunteers have joined militias, paramilitary units, guerrilla outfits, and regular armies, navies, and air forces. They have fought in insurgencies, civil wars, wars of secession, and state-on-state wars. But while contexts may differ, certain mechanisms recur.[12] In fact, foreign volunteers of different generations share a number of characteristics. These commonalities are the subject of this book.

Definitions and (Porous) Boundaries

Foreign volunteers leave their country of nationality or residence and take part in a conflict abroad on the basis of a personal decision, without being sent by their government and not primarily for material gain.[13] Such a definition excludes from our discussion groups that have been labeled as "volunteers" but were, in fact, instruments of government policy.[14] During the Spanish Civil War the Italian dictator Benito Mussolini sent tens of thousands of Fascist Blackshirts and conscripts from the Italian armed forces to fight against the Spanish Republic. Part of this force, known as the Corpo Truppe Volontarie, included individuals who had volunteered for service in Spain. Yet, importantly for our purpose, it was commanded by an active general in the Italian army and was supplied, paid for, and directed by the Italian foreign ministry's Ufficio Spagna in Rome.[15] Another military force that will not be discussed is the Spanish Blue Division, which was sent by the government of the dictator Francisco Franco to fight alongside Nazi Germany against the Soviet Union in the Second World War. Like Fascist Italy's force in Spain, the Blue Division was composed of Falangist volunteers, but it also included men from the Spanish army, some of whom were coerced to "volunteer." The Spanish government, which formed this 17,000-strong force in summer 1941, decided to disband it in 1944 once the Nazi war effort seemed to be heading toward defeat. Between 400 and 500 Spaniards decided to stay on and fight alongside the Germans, and some of these took part in the final defense of Berlin in 1945.[16] Those who chose to remain until the bitter end, even after the Spanish government pulled out, can be considered foreign volunteers.

To further clarify who is included and who is excluded from our definition of foreign volunteers, let us consider two cases of US citizens who served abroad during the Second World War. The group of seventy-four American pilots and support staff who fought alongside Chinese nationalist forces against Japan from 1941 onward are not discussed in this book. They were, for all intents and purposes, seconded by President Franklin D. Roosevelt, who signed an order on 15 April 1941, permitting American military pilots to join General Claire Chennault, who was forming an American Volunteer Group in China. These men, who came to be known as the Flying Tigers, were an instrument of the foreign policy of the United States.[17] In contrast, those American airmen who joined Britain's Royal Air Force (RAF) between 1939 and 1941, with some of them going on to form the Eagle Squadrons, are discussed. As we will see, many of these volunteers traveled to Canada to enlist, sometimes surreptitiously, and were at risk of losing their American citizenship for doing so.

The definition, and the book in general, speaks of "foreign volunteers" rather than "foreign fighters" for two main reasons. First, historically, the term "volunteers" has been used widely and fairly consistently since the phenomenon emerged. Former US president John Quincy Adams described the Marquis de Lafayette as "a volunteer" in his 1846 Lives of Celebrated Statesmen.[18] Thomas Gordon, one of the foreigners who fought for the independence of Greece in the 1820s, used the phrase "European volunteers" in his 1832 History of the Greek Revolution.[19] The foreigners who joined the International Brigades and fought in the Spanish Civil War were invariably referred to as volunteers.

Of course, the phrase is not unique to texts in English and is prevalent in other languages as well. For instance, the thousands of foreigners who offered their services to France in 1914, following the outbreak of the First World War, were referred to at the time as *volontaires*.[20] The foreigners who enlisted in the Nazi Party's Waffen SS and fought for Germany in the Second World War were called *Freiwilligen*.[21] The foreigners who fought for Israel's independence in the war of 1948 were called *mitnadvim*. The same can be said about equivalent examples in Finnish, Italian, Serbo-Croatian, Spanish, and other languages. The term "foreign fighter" is fairly novel. It has come into use mainly in the context of Muslims who have fought transnationally in various conflicts since the 1980s.[22] Therefore, I have used "foreign fighter" only where such recent conflicts are discussed.

A second reason the term "foreign volunteers" has been selected is to emphasize the importance of voluntarism as one of the phenomenon's key attributes. These were individuals who *chose* to take part in a conflict, to take orders, to put their lives at risk, and, potentially, to take the lives of others. It is this decision that marks them out from the millions throughout history who were conscripted into service or did so to earn their wages. The distinctiveness of the foreign volunteers' situation is captured well in a conversation that Flora Sandes, a British woman who joined the Serbian army in the First World War, had with her Serbian comrades: "like the Turks they say, 'to die for your country is not to die'; but to die for someone else's country they thought to be something extra special."[23]

Importantly, not all the foreigners who enlisted to support the war effort of another country or political entity were necessarily "fighters." Lord Byron, for instance, died of illness in Greece in 1824 before he got a chance to do any fighting. Armed foreign volunteers have much in common with non-military transnational volunteers who travel to war-stricken areas to provide aid or medical relief. Combatant and non-combatant volunteers often respond to the same international crises. For example, the Afghan struggle against the Soviet Union in the 1980s was the conflict where the young Saudi extremist Osama bin Laden made his international debut. The son of a construction industry magnate, bin Laden's initial aim was to build tunnels for the benefit of the Afghan mujahideen using the bulldozers, trucks, and equipment he had procured. In fact, the vast majority of foreigners who sought to help the Afghan cause in the second half of the 1980s ended up staying in neighboring Pakistan and working as refugee-camp teachers, cooks, accountants, and doctors.[24] Although our focus here is on foreigners who engaged in fighting, the boundary between armed and non-military volunteers has sometimes been a porous one. The two categories share a number of characteristics, and there have been a few foreign aid workers who decided to take up arms. As "Medya" (nom de guerre), a German doctor from Hamburg who ran a health care program for the Kurdish Workers' Party (PKK) in northern Iraq, explained to a TV crew in 2007, "if the moment comes when I or my patients are attacked, I would not hesitate to reach for my gun."[25]

While the book uses a fairly broad definition of foreign volunteers, it is not concerned with the buying and selling of military service as such. Mercenarism and foreign volunteering are separate phenomena, although the boundary between the two has not always been so distinct. Indeed, the world that historians see is rarely divided into neat categories. To begin with, except for a small minority who paid their own way, foreign volunteers normally received some sort of pay as well as food and accommodation during their military service. The key difference is whether material gain was a central motivation for enlisting and for remaining in service. The UN International Convention against the Recruitment, Use, Financing and Training of Mercenaries defined the latter as

> motivated to take part in the hostilities essentially by the desire for private gain and, in fact, is promised, by or on behalf of a party to the conflict, material compensation substantially in excess of that promised or paid to combatants of similar rank and functions in the armed forces of that party.[26]

The UN definition, and international law concerning mercenaries in general, has been criticized from various perspectives.[27] However, it does clearly demarcate those who are not foreign volunteers. Even so, the historical record has produced a number of gray-zone cases. For instance, historians disagree on whether the European recruits in Simón Bolívar's army and navy during Latin America's struggle for independence from Spain in the late 1810s and early 1820s should be considered volunteers, adventurers, or mercenaries.[28] How should foreigners such as these be considered, when some of them were promised land in return for their service? In this book I discuss borderline cases like Bolívar's foreign soldiers and sailors but ignore blatant examples of mercenarism such as Mike Hoare, a former British army officer who received handsome amounts of money for his military services in Congo and in other conflicts in the 1960s and 1970s.[29]

The French Foreign Legion and its Spanish equivalent are largely excluded from the following chapters. Recruits who joined the Legion were motivated by a blend of practical need, adventurism, idealism, and desperation.[30] However, once they enlisted, they often had little or no influence over where they were posted or which conflicts they would fight in. Foreign Legionnaires are considered in a couple of cases where there was a clear ideological component in their enlistment considerations—namely, the Italian *garibaldini* who joined the French Foreign Legion following the outbreak of the First World War, and the right-wing foreigners who went to Spain during the civil war and joined the so-called *Tercio* to support the Nationalists led by General Franco. Volunteers such as these were funneled by their hosts into Foreign Legion units and did not always fit the legionnaire mold. As Peppino

Garibaldi, the leader of the Italian volunteers in France, told a French general who was dissatisfied with the new recruits' lack of discipline in autumn 1914: "What do you want, General? We are volunteers, not *legionnaires*."[31]

This book does not concern itself with private military or security contractors, an industry that has begun to thrive since the end of the Cold War. While such companies often participate in armed conflicts and operate transnationally, their raison d'être is commercial. However, there have been a few cases of individual contractors who temporarily chose to become foreign volunteers. For example, John "Olly" Scarrott, a former British Royal Marine, worked as a military advisor on oil and gas tankers in the Persian Gulf during the Iran-Iraq War (1980–1988). In 1991 he left his job on an off-shore oil rig near Nigeria to fight in Croatia's war of independence. In early 1992, after a number of months in Croatia, he decided to leave, returning to his previous line of employment.[32] Another individual who served as a foreign volunteer is Rob Krott, who fought in the Yugoslav Wars (1991–1995) in between jobs as a military or security contractor. A former officer in the US Army, Krott volunteered to serve in a Croatian military unit in 1992 and worked for an American defense contractor in Somalia later that year before returning to a second voluntary "tour" in the Balkans, this time in Bosnia-Herzegovina, in spring 1993. In the 2000s he worked as a security contractor in Iraq.[33] Very rarely has foreign volunteering been a decades-long commitment, wherein individuals fought voluntarily in one conflict after another. The phenomenon is better understood as a phase in the lives of certain people.

A final qualification needs to be made about the omission of colonial troops from our discussion. Colonial troops have played a very large and significant role in the world's military history in the nineteenth and twentieth centuries. For instance, the Indian army, which formed part of the British Empire's military until 1947, committed millions of troops to the Allied war effort over the course of the First and the Second World Wars. Moreover, recruitment in the Indian army was, by and large, voluntary. Service in the Indian army could be considered transnational insofar as multinational empires are understood as transnational entities.[34] But such service was not "foreign." Colonial recruits where serving "their" state, even if they did not identify with all its symbols or perceived themselves to be part of a different nation. The colonial state claimed to hold what the sociologist Max Weber called the monopoly over the legitimate use of physical force within its territory.[35] The enlistment of indigenous troops met the colonial state's expectations and served its domestic and international needs. As we shall see, foreign volunteers who serve in a foreign conflict, without being sent by their government, defy international norms and expectations.

Themes and Patterns

The chapters of this book are structured thematically. Each one highlights a different facet of the phenomenon. During the period under examination, the world has seen tens if not hundreds of thousands of foreign volunteers. This book does not presume to recount every story or even trace every conflict in which foreign volunteers fought. Instead, it focuses on various individuals and groups of foreign volunteers who participated in different conflicts and who exemplify some sort of common trait or broader theme. Readers may well recognize other historical or contemporary figures in the examples I have chosen to discuss.

Chapter 1 surveys the changes in recruitment practices since the Middle Ages and explains why the late eighteenth century, and particularly the French Revolutionary Wars, provide a starting point for the modern-day foreign volunteer phenomenon. Chapter 2 examines the role ideologies have played in attracting foreign volunteers to various conflicts. It also offers a periodization that divides the conflicts that are discussed in detail in subsequent chapters into three ideological "waves." Personal motivations for volunteering and the social and cultural backgrounds from which prospective volunteers emerged are the topic of Chapter 3. When describing the volunteers' motivations and experiences, I have tried to use their own words whenever possible.

Chapter 4 divides foreign volunteers into different categories based on how they positioned themselves in relation to their home state. The chapter argues that a classification such as this serves as a more effective analytical tool than divisions based on motivations. Chapter 5 examines the responses of home states and the international community to the foreign volunteer phenomenon. It surveys various historical and contemporary legislative attempts to restrain the flow of foreign volunteers. It also illustrates how transnational networks that were set up by the volunteers and their supporters have been key in helping foreigners bypass national and international restrictions. Chapter 6 assesses the military significance and effectiveness of various cohorts of foreign volunteers. It illustrates how their most consistent contribution is often in the realms of politics and propaganda rather than on the battlefield. The negative sides of the phenomenon such as criminality and harmful encounters between volunteers and their hosts are discussed in Chapter 7. The same chapter also contends with various hypotheses about the role of foreign military service in radicalizing volunteers, including in the context of international terrorism. Chapter 8 deals with memory. It identifies certain recurring ways in which foreign volunteers remembered and positioned themselves in relation to foreign volunteers of previous generations.

The book should not be read as a manual for practitioners, hoping to stem the phenomenon. Nor should it be seen as a guidebook for prospective foreign volunteers. Readers who are contemplating voluntary, "non-state" military service abroad should either stop reading now or skip forward to Chapter 8, where the discussion of the myths that have evolved around certain individuals and groups of foreign volunteers may provide some encouragement. My goal is to uncover a number of patterns that would help us understand a long-standing facet of war that has manifested itself in different places and at different times. The examples in this book try to reflect this global phenomenon's geographic and temporal breadth, though inevitably some regions received more attention than others.[36] While by now nearly every country on earth has produced foreign volunteers, it was in Europe and North America where the link between nationality, citizenship, and military service first emerged as a norm, and therefore it is there that our story begins.

* 1 *

Only a Nation in Arms?

"Foreigners" in Military Service before 1815

When Foreign Soldiers Were the Norm

January 1574. Roger Williams, a freelancer in the true sense of the word—a lancer by profession and a soldier of fortune—was trying to make his way back to England from Germany, having run out of money. Williams recounts that in the town of Lier, located in what is now northern Belgium, a Spanish commander

> requested me earnestly to try his courtesy in the Spanish army, assuring me to depart when [it] pleased me. Having spent all my crowns, and being loath to return into England without seeing something, I promised to stay. Also, in those days there was no dispute betwixt Her Majesty [Queen Elizabeth I] and the Spanish king, to my knowledge. This was the manner and the first hour that I entered into the Spanish service.[1]

Considering the norms that were prevalent at the time, there was nothing exceptional about Williams's decision. Not even the fact that in the two years previous, he had fought for the Dutch, the enemies of the Spanish in the Eighty Years' War, also known as the Dutch War of Independence (1560s–1648). English, Scottish, Walloon, German, and Dutch soldiers were found in the ranks of the Spanish army, as well as on the opposing side. At one point in this prolonged conflict, French-born René Descartes, who later became one of the fathers of modern philosophy, served in the army of the Dutch States.

But let us return to Williams. In 1578 he switched sides again, fighting for the Dutch, first as a mercenary and later in an English contingent sent by Queen Elizabeth to aid the Netherlanders. In fact, Williams was able to move quite freely between the service of foreign rulers and that of his sovereign. In 1588–1589 he took part in preparations to repel the Spanish Armada, which was expected to land in England, and in the 1590s he fought in the wars of religion in France.[2] A Welshman known for his valor in combat—he may or may not have been the inspiration for the character of Fluellen in Shakespeare's *Henry V*—Williams also wrote extensively about his own experiences and about military theory. "Dutie, honor & welth, makes men follow the wars," he argued, adding that there was "no disgrace for a poore Gentleman that lives by warres to serve any estate that is in league with his owne."[3] He admitted in retrospect that for a while he had been unhappy fighting for the Dutch and longed to quit their service, but "Hungry dogs must follow such that gives them bread."[4]

This example from the late sixteenth century sits well with the classic historical narrative on the evolution of military recruitment in Europe from the late Middle Ages to the present. According to this narrative, the medieval West was dominated by "feudalism": knights, who were feudal vassals, served their lords for forty days each year in return for fiefs.[5] With the breakdown of the feudal system, knights who had amassed substantial amounts of money were able to pay rent to their lord instead of providing military services. This in turn led monarchs increasingly to rely on hired soldiers (derived from the Italian word *"soldi"*—money) to fight their wars. There were social and economic advantages to the employment of mercenaries. Hiring outsiders instead of using one's own vassals guaranteed the continuity of agriculture, ensured the profitability of landed estates, avoided social disruption and minimized wastage of manpower.[6]

However, there were also problems inherent in this system of military recruitment. The fourteenth century saw the emergence of the phenomenon of the free companies—large organized bands that effectively replaced the individual recruitment of mercenaries. The Great Company, a band approximately

10,000 strong and composed of soldiers from different countries, ran what one modern-day historian has labeled "a protection racket" across the Italian peninsula from the 1330s to the 1350s.[7] This unruly system became more stable and reliable with the rise of the *condottieri* toward the end of the century. These more institutionalized warlords were persuaded to put down roots in the states they served by receiving concessions of land in return for military service. A *condottiero* had an interest in preserving his name and reputation by achieving victories. But it would be ill-advised for such victories to be too crushing, lest there would be no need for further campaigns. Moreover, trained men were expensive and difficult to replace, and *condottieri* were want to waste them. Hence, when the great Italian historian, politician, and philosopher Niccolò Machiavelli (1469–1527) described the military conflicts of the fifteenth century in his *History of Florence and of the Affairs of Italy,* he famously lamented that "the practice of arms fell into such a state of decay, that wars were commenced without fear, continued without danger, and concluded without loss."[8]

The fifteenth, sixteenth, and seventeenth centuries saw a gradual rise in the powers of the state, the creation of standing armies, and the persistence of international recruitment. Louis XI of France (1423–1483) was probably the first to maintain a large standing army in peacetime where troops were paid regularly, even when they did not fight. Approximately half of his forces were recruited from beyond the frontiers of France. At this stage, however, the king still did not have a monopoly over raising armies. Power struggles between the crown and the aristocracy continued until the conclusion of the Thirty Years' War in 1648. In the second half of the seventeenth century, European princes generally acquired sufficient control of their territorial resources to maintain standing armies on a continuing basis. The central administration of armies took hold with the establishment of military secretaries in France (1635), Austria (1650), Britain (1661), and Piedmont (1717). Gradually, war became an activity regulated by the state.[9]

People from across the social range joined the ranks of the military for a variety of reasons. Many men chose the army because it offered employment at a time when civilian life did not. In 1572 the Venetian military engineer, General Giulio Savorgnan, explained that men enlisted

> to escape from being craftsmen [or] working in a shop; to avoid a
> criminal sentence; to see new things; to pursue honour (though

these are very few) . . . all in the hope of having enough to live on and a bit over for shoes, or some other trifle that will make life supportable.[10]

Others enlisted out of a sense of adventure or because they desired a change of scenery. Sir James Turner (1615–1686), who as a teenager fought for both Denmark and Sweden during the 1630s, confessed that he went to fight because "a restless desire [had] enter'd my mind to be, if not an actor, at least a spectator of those warrs which at the time made so much noise over all the world."[11] A customary form of recruitment during the Thirty Years' War was for victorious commanders to fill their depleted ranks with recruits from the defeated armies of their opponents. This should come as no surprise in a war in which the soldiers were mercenaries and the officers entrepreneurs.[12]

Although their view has recently been challenged, key classic-narrative historians agree about the lack of esprit de corps among European forces during this period.[13] John Childs has argued that "passion was absent from the battlefield and war was limited in every sense. Conflicts were between the leaders of dynasties without much reference to society. . . . Not only were the resources of the sixteenth-century state finite but the number of mercenaries in western Europe was strictly limited."[14] According to Michael Howard, one of the doyens of military history in Britain, "Whatever the rationale of wars during this period, whether disputes over inheritance or, as they became during the latter part of the sixteenth century, conflicts of religious belief, they were carried on by a largely international class of contractors on a purely commercial basis."[15] Similarly, Geoffrey Parker's influential book *The Military Revolution* claimed that the "combination of heterogeneous methods of recruiting, high wastage rates, and considerable mobility within the ranks soon destroyed any sense of corporate identity among the individual formations of every early modern army."[16]

A combination of profound changes led to a massive growth in the size of European armies in the eighteenth century. Developments in agriculture increased the availability of food for troops and fodder for their animals. Improved roads made it easier to transport forces. And, finally, European rulers, who devoted more than 70 percent of their revenues to war, needed to keep up with each other. Once one monarch enlarged his or her army, the others had to follow suit. The Habsburg army grew from approximately 50,000 in

1690 to 200,000 in 1778. The Prussian army, 39,000 strong in 1710, numbered 160,000 in 1778. In Russia the army expanded from 90,000 in the late seventeenth century to a staggering 500,000 in 1789. The growth in the size of European armies outstripped the increase in Europe's population.[17]

This monumental growth led rulers to introduce early forms of conscription. Peter the Great of Russia (1672–1725) paved the way, and similar policies were implemented in Prussia and other Germanic states. Local magistrates were entrusted with selecting conscripts and were often tempted to enroll personal enemies and to rid themselves of society's undesirables at the army's expense. The Prussian monarchy specifically exempted valuable tax payers—merchants, artisans, manufacturers—and their sons from military service. Furthermore, a number of cities such as Berlin were relieved of the burden of military service. This was done to impose the least possible strain on the country's fragile economy. Consequently, historians have tended to view eighteenth-century soldiers as a despised group on the periphery of society.[18] Armies, in the words of Frederick the Great of Prussia (1712–1786), were "for the most part composed of the dregs of society—sluggards, rakes, debauchees, rioters, undutiful sons, and the like, who have little attachment to their masters or concern about them as do foreigners."[19]

Desertion was a perennial problem for eighteenth-century armies. In the War of the Spanish Succession (1701–1714) one in four French soldiers deserted from the ranks. During the Seven Years' War (1756–1763), some 80,000 men absconded from the Prussian army, 70,000 from the French, and 62,000 from the Austrian. Although penalties were severe—in France every third deserter who was recaptured was shot—it often proved difficult to keep press-ganged soldiers, many of whom were recruited from the peasantry, under arms.[20]

The unreliability of conscripts and the desire not to impose unnecessary strains on the local economy ensured the persistence of the recruitment of foreigners. Poorer, smaller German states rented their manpower to richer countries. During the American Revolution, for instance, the Hessian states in Germany furnished 30,000 men for the British army. Foreign troops entering the French Royal Army were organized into Swiss, German, Irish, Italian, and Liégeois regiments under their own officers. The words of command were usually given in the soldiers' native language. Officers too were not restricted to serving in the army of their country of origin. Because European armies were organized along roughly equivalent lines, used very similar tactics and weapons, and offered comparable rewards, good officers could take up a commission with almost any army. Throughout the eighteenth century the number of foreigners in the armies of Europe fluctuated and varied from country to country. From lows of around 15 percent, it could reach as high as more than 50 percent, especially in times of peace. The advantages of employing foreigners were so noticeable that, as one French general remarked, "each foreign soldier was worth three men, one more for France, one less for the enemy, and one Frenchman left to pay taxes."[21]

In the second half of the eighteenth century the ideas of the Enlightenment began to influence first the theory and eventually the practice of military recruitment. The witty and hugely influential Voltaire mocked the practice of press-ganging unsuspecting recruits into military service. The protagonist of his *Candide* (1759) is tricked, handcuffed, and carried away into a regiment of the king of the Bulgarians.[22] In his *Essai général de tactique* (1772), the general and military theorist Jacques Antoine Hippolyte, Comte de Guibert, stressed the importance of the relationship between war and politics. He argued that the nature and composition of an army—in his case, the French—should be rooted in the nation's customs and constitution instead of reflecting those of its neighbors. Drawing on the idealized classical models of ancient Sparta and Rome, he advocated the creation of an army of citizens, members of a nation whose patriotism would motivate them in action. At the heart of the problem for Guibert was the inability of governments "to compose our armies of citizens, men who have the zeal for the service, or soldiers, not merely for the sake of gain." Instead of making "honour their reward," the state "pays them with money alone."[23]

Such ideas began to make their mark across the Atlantic. When the conflict between Britain and its thirteen colonies in North America erupted in April 1775, revolutionary leaders in Congress and elsewhere wanted their struggle to be based on citizen-soldiers joining militia units to defend their homes. One of the grievances enumerated in the Declaration of Independence of 4 July 1776, was the use King George III had made of troops hired in Germany: "He is at this time transporting large Armies of foreign Mercenaries to compleat the works of death, desolation and tyranny."[24]

The classic narrative, however, identifies the wars that followed the French Revolution as the crucial turning point in European recruitment practices. Revolutionary leaders spoke in terms of creating a "national" army, along the lines suggested by Guibert, as early as autumn 1789.[25] In reality, though, the Revolution created a considerable degree of disorganization and uncertainty in the ranks of the French army. Desertion among the rank and file gained momentum and was accompanied by a growing number of officers who chose to emigrate. A series of small risings culminated in a large mutiny involving three regiments (two French and one Swiss), in Nancy in summer 1790. The mutiny was repressed by force. However, the National Assembly embarked on a series of reforms to strengthen the cohesion of the army, to increase its reliability, and to better defend France from external enemies. In 1791 the authorities called on young men to volunteer for military service. The call to arms initially met with enthusiasm. Some 100,000 volunteers enlisted. Changes were also introduced in the officer corps, where meritocracy was to replace social rank and privilege. Importantly for our purpose, the leadership sought to end France's high dependency on foreign mercenaries. From mid-1791 foreign recruitment all but ceased. The National Assembly abolished the distinctions between French and foreign regiments, enrolling thousands of French citizens as soldiers into the latter.[26]

Xenophobia, as one historian recently pointed out, was "a feature of the Revolution from its earliest moments."[27] The anti-mercenary and anti-foreigner mood found its way into the words of the *Chant de guerre pour l'Armée du Rhin,* known to us today as the *Marseillaise.* Written in April 1792, at a time when France found itself at war with Austria and Prussia, the lyrics include the following invocation:

> To arms, citizens,
> Form your battalions,
> Let's march, let's march!
> Let the impure blood
> Water our furrows!
>
> What! Foreign cohorts
> Would make the law in our homes!
> What! These mercenary phalanxes
> Would strike down our proud warriors![28]

This anti-mercenary atmosphere reached a climax after the king's Swiss Guard was accused of opening fire on a Paris crowd at the Tuileries on 10 August 1792. In the mayhem that ensued, 650 out of the 900 Swiss troops present were killed and a further 100 were wounded. Subsequently, the Legislative Assembly decided to disband the Swiss regiments on 20 August.[29]

Following the outbreak of war, the leadership declared that *la patrie* was in danger. A second call was issued for more volunteers. When the number of people offering to serve proved insufficient, France's Jacobin leaders decided to mobilize the entire population—the famous *levée en masse* of August 1793:

> From this moment until that when the enemy is driven from the territory of the republic, every Frenchman is commandeered *(en requisition permanente)* for the needs of the armies. Young men will go to the front: married men will forge arms, and carry food: women will make tents and clothing, and work in hospitals: children will turn old linen into bandages: old men will be carried into the squares to rouse the courage of the combatants, and to teach the hatred of kings, and republican unity.[30]

In assessing the importance of this last development, Carl von Clausewitz (1780–1831), one of—if not the—most important military theorists of the modern age, set the tone for future generations of historians. Clausewitz was a Prussian officer and a veteran of the Napoleonic Wars. His *On War,* a book edited and published posthumously by his wife in 1832, argued that "in 1793 a force appeared that beggared all imagination."[31] This force was nationalism. Hitherto, war had been "solely the concern of the government," a cabinet affair "in which the inhabitants were not expected to show any particular interest." The French Revolution, which sought to turn Frenchmen into citizens and involve them politically and emotionally in the fate of the state, created a profound change: "Suddenly war again became the business of the people—a people of thirty millions, all of whom considered themselves to be citizens."[32] France, in other words, had become a nation in arms. The subsequent Napoleonic Wars, especially the stubborn Spanish resistance against French rule, the Russian campaign of 1812, and the Prussian-led victories of 1813, were manifestations of the spreading force of national mobilization across Europe. They revealed "what an enormous contribution the heart and temper of a nation can make to the sum total of its politics, war potential and fighting strength."[33] Echoes of this Clausewitzian notion can be found throughout the historiography on military recruitment.[34]

There is one more aspect of the classic narrative that is important to emphasize for our purpose. As George Mosse and others have pointed out, following the French Revolution, military service became far more respectable than it had been under the ancien régime. Recruits were no longer drawn from the margins of society. As citizens who were also an integral part of the local community—the sons, brothers, or neighbors of a cross-section of society—they were fêted by the republic. Those who died defending France were celebrated by nationalist propaganda. From the Napoleonic Wars onward, soldiers' final resting places became "shrines of national worship," and monuments erected in their memory were positioned in central and public spaces. This custom became prevalent across Europe, the United States, and in various other parts of the world during the nineteenth century, reaching its peak around the First and Second World Wars.[35] With this nationalization of military service in mind, it is perhaps no wonder that the mercenaries of the eighteenth century gave way to the citizen-soldiers of the nineteenth and twentieth centuries.

As one might expect, revisionist historians in recent decades have been picking away at this classic narrative. For instance, the assertion made by Clausewitz about 30 million Frenchmen all considering themselves to be citizens has been called into question. A number of studies have shown that the rural and urban recruits of the 1790s tended to be poor. Many of them were persuaded to enlist by economic need and necessity rather than patriotism. In other words, their motivations were similar to those that had for centuries underpinned recruitment in the French Royal Army and in professional armies across Europe. Furthermore, during the Revolutionary Wars the French army continued to be plagued by desertion, much like the forces of the ancien régime had been. Fewer than half the peasants who were liable for military service reported to duty, and 200,000 men fled service annually.[36] Hence the divisions, which apparently separated eighteenth-century Europe from what followed the upheaval of July 1789, did not alter every single aspect of military organization.[37]

The revolutionary principle, according to which national armies ought to do the fighting in national wars, became much more blurred in the wake of French expansion under Napoleon. Increasingly, regiments were raised among the inhabitants of occupied lands. More than a third of the soldiers in Napoleon's Russian campaign were not Frenchmen. Instead, they were either conscripts from territories that had been annexed by the French Empire or came from contingents that Napoleon's allies were expected to deliver. Hence, a large portion of the Grande Armée was drawn from the populations of Belgium, Poland, Italy, Switzerland, and the German states. The letters, diaries, and memoirs written by some of the veterans of Napoleon's wars suggest that a sense of national belonging was only one of many considerations influencing their decisions.[38]

Ironically, Clausewitz himself knew a thing or two about the relative ease of joining the army of another country during the Napoleonic Wars. Along with other middle-ranking officers, he had tried to push for reforms in the Prussian army after its defeat at the Battle of Jena-Auerstädt (1806). He wanted Prussia to adopt measures in response to the new French way of war, but his hopes were dashed by the Prussian king's hesitation. In 1812 Prussia was forced to assign forces to the army Napoleon was assembling for the invasion of Russia. Consequently, Clausewitz and a few of his comrades resigned their commission and joined the Russian army. He served as staff officer with several Russian commands, but because he did not speak Russian, his contribution was limited. In 1813, after Napoleon's catastrophic retreat, Clausewitz returned to Prussian service and later became the head of the military academy in Berlin.[39]

Another critique of the narrative that sees the Revolutionary and Napoleonic Wars as a watershed in military history relates to the durability of compulsory universal military service that was first imposed in 1793.[40] After 1815 Prussia was the only European state to retain conscription without exemption or substitution. As recent studies have shown, the levée en masse in France and attempts to emulate it elsewhere proved to be an episodic rather than a permanent phenomenon.[41] Following the restoration of the Bourbon monarchy, the French army was reconstructed into a smaller professional force. It included two Swiss infantry regiments of the Royal Guard and the Hohenlohe Regiment that was also comprised of foreigners. When these regiments were disbanded in 1830–1831 they were replaced by the newly established Légion Étrangère, the French Foreign Legion.[42] In short, despite the political upheaval and the undeniable changes wrought by the Revolution of 1789, there were important lines of continuity linking early-modern, revolutionary, and even post-revolutionary warfare. Although it underwent changes in volume and in form, the enlistment of foreigners in European armies, which had been the norm for centuries, did not cease overnight.

But there is an important element missing from the image we have constructed thus far, which has emphasized the buying, selling, and conscripting of military service: the role of ideas in encouraging "foreign" enlistment. Before we examine how the Revolutionary and Napoleonic Wars helped to shape the phenomenon of foreign volunteering for the generations that followed, let us see what motivated their medieval and early-modern forerunners. Here too a complex picture that combines continuity and change emerges.

Forerunners: Fighting Abroad for Ideals before the Age of Revolution

Not all of the foreign soldiers in the centuries leading up to the French Revolution enlisted to fight for the sake of material gain or purely because of adventurism. Modern foreign volunteers are often associated with ideology. Which principles motivated their earlier predecessors? A good starting point in our search for distant historical precedents for the modern phenomenon of foreign volunteers would be the Crusades. The First Crusade of 1096–1099, before crusading became an institutionalized form of military service, provides an interesting case for comparison. There are a few similarities but also a number of key differences.

In November 1095 Pope Urban II gave a speech in the town of Clermont, calling on the Christian West to come to the assistance of the Byzantine Empire in waging holy war against the enemies of Christianity in the Levant. This call, most likely echoed by preachers and word of mouth across western and central Europe, prompted thousands to volunteer their services in large assemblies. Soon a powerful and quite spontaneous movement was under way. The participants in the First Crusade came from various kingdoms and principalities that today comprise France, Germany, Belgium, Italy, and England. Not led by a king of their own, the crusaders—or *cruces-ignati,* as they were eventually known—assumed that they were going to campaign under the command of the Byzantine emperor, Alexius I Comnenus. However, the Byzantines were unsettled by these unexpectedly large forces, which arrived in a number of waves. Therefore, Alexius made them swear an oath to return the territories they would capture to him and then used every means at his disposal to persuade each contingent to cross the Bosphorus into Asia. The crusaders proceeded eastward and then southward, fighting their way. After Alexius's troops failed to help them during the Siege of Antioch, the crusaders saw themselves as liberated from their oaths to him. They continued as an independent force and eventually established four new Christian principalities in the Levant, the most famous of which was the Kingdom of Jerusalem. In terms of size, around 100,000 people set out from Europe in 1096. However, numbers dwindled very significantly on the way. From an army of around 70,000 men (and women), including about 7,000 knights, that gathered at Nicaea in Asia Minor in 1097, only some 12,000 to 14,000 reached Jerusalem in 1099.[43]

The motivations of the crusaders, as far as we can tell from the available sources, were complex. As one might expect, religious reasons played a central role. These could include the guilt of a repentant sinner, the desire for an adequate penitential endeavor, or the fear of the prospect of hell. Crusaders fought for God in return for a promise of salvation. The council of the Church at Clermont, which formally launched the Crusade in November 1095, decreed: "Whoever for devotion alone, not to gain honour or money, goes to Jerusalem to liberate the Church of God can substitute this journey for all penance." The charter of one departing crusader in 1096 stated that he was going to fight "for God against pagans and Saracens [Muslims]."[44] There was a rhetoric of popular participation in crusading that makes the Crusades rather different from standard wars in medieval Europe. In some instances, those who volunteered were expected to have the crosses that had been distributed to them sewn on to their clothes at once, and to go on wearing them until they came home having fulfilled their vows.[45]

As plunder was an integral part of medieval warfare, the Crusades have also been described over the centuries as plundering expeditions. Of course, material and spiritual motivations are not mutually exclusive. However, while a few families profited substantially, it should be noted that few of the crusaders who returned to the West seem to have brought valuable treasures back with them. Furthermore, crusading cost the families of those who took the cross a tremendous amount in financial terms.

Another interpretation that was put forward during the twentieth century saw the Crusades as a colonial movement, the opening act in the "expansion of Europe." However, as Jonathan Riley-Smith has shown, the majority of crusaders returned to Europe following the conquest of Jerusalem. Only about one in nine or ten settled in the East. Most of the settlers who lived in the newly established Christian principalities were not crusaders in the narrow sense of the term because they did not participate in the initial conquest.[46]

Adventure and opportunity cannot be dismissed as motivating factors. Crusading was inconvenient, dangerous, and expensive. The majority of western Europe's population stayed put and did not set out toward the Middle East. Those who took the cross of their own accord must have been prepared to leave what we would now call their comfort zone. It is worth noting, however, that prosopographic studies of the Crusades have revealed that many early crusaders were clustered into certain kindred groups: they tended to come from the same locales and were often related to one another.[47] Hence, obligations toward social superiors and peers should also be added to the list of motivations.

On the level of individuals, the coexistence of ideological zeal (in this case, religious belief), opportunity, adventure, and social networks in the motivations and considerations of crusaders is not dissimilar to the reasons that prompted foreign volunteers in the modern era. However, the social, cultural, and political context in which the crusaders lived was fundamentally different. To begin with, those Europeans who took the cross in the late eleventh century were not citizens of a nation-state. Instead, they were vassals who had taken an oath of fealty to their lords. In the words of Christopher Tyerman, "In a society in which in many regions the bulk of the population were bound to landlords by servile tenure, only freemen could legitimately take the cross."[48] Serfs were not freemen. If a nobleman decided to join the Crusade, a retinue of knights and infantry would follow him, either because they were obliged to do so or—if they were free to make up their minds—because they believed that this would best serve their interests. As Thomas Asbridge aptly put it, this created "a domino effect, whereby for every noble who took the cross a chain reaction was initiated."[49] In other words, not everyone who went on Crusade was, strictly speaking, a volunteer.

A second distinction is that the crusader cause was not exactly foreign to European rulers. The First Crusade was instigated by the pope (and the rivalry for political supremacy between the papacy and the lay rulers was a central tenet of medieval European history). While no monarch participated in the First Crusade, Philip I of France (1052–1108) showed enthusiasm for it.[50] And it was not long before kings assumed the leadership. Richard I of England—Richard the Lionheart—famously led the Third Crusade (1189–1192), which sought to salvage what remained of the crusader lands in the Levant after the victories of Salah al-Din (also known as Saladin).

A third important distinction revolves around what the population was and was not expected to do. Compared to modern states, the eleventh-century kingdoms and principalities of Europe were weak and fragmented political entities. Philip I, for instance, struggled to control some of the areas near Paris, his capital, let alone manifest his will throughout the whole of what we now know as France. There were no full-time standing armies. Even knights had other roles as lords, vassals, landholders, or farmers. Furthermore, knights could own possessions for which homage and fidelity were owed to different lords, in diverse locales. Their political obligations were crisscrossed rather than strictly hierarchical (and there was always potential for these connections to conflict with one another).[51] Monarchs had neither the means, the administrative capacity, nor what Max Weber called the monopoly over the use of force, to exercise the level of control that modern states have over their citizens. To paraphrase the classic-narrative sociologist Charles Tilly, the Crusades took place before the state made war and war made the state.[52] A *levée en masse* such as the one attempted by the Jacobin leaders of France in 1793 would have been unimaginable in the 1090s. As we shall see later on, several modern states passed laws prohibiting their citizens from serving in foreign armed forces. Such legal restrictions were still a thing of the distant future when the First Crusade got under way.

Of course, people fought at least partially for the sake of ideals in other medieval and early-modern wars. Guy Fawkes (1570–1606), because of whom British children light bonfires and fireworks every 5 November to this day, is one such example. A Yorkshire-born Catholic, Fawkes left England for the Low Countries in 1592, joined the armies of Catholic Spain, and fought against the Protestant Dutch. Fawkes was a soldier by profession. One acquaintance described him as "a man highly skilled in matters of war."[53] However, despite his reputation for bravery, Fawkes only rose to the rank of ensign. In 1603 he tried, unsuccessfully, to convince the Spanish court to set out on a new military venture in aid of English Catholics. After this point, to use our contemporary terminology, Fawkes crossed the threshold between a "foreign fighter" and a "terrorist." Wishing to topple Protestant rule in England, he soon became involved in the notorious gunpowder plot: the plan to blow up the Palace of Westminster, where parliament sat, on 5 November 1605. Arrested while emerging from a cellar where thirty-six barrels of gunpowder were found, Fawkes was put on trial and subsequently executed.[54]

It is worth noting that service with Spanish forces, while certainly unusual for an Englishman at the time, was by no means unique. As we have seen, Sir Roger Williams fought for the Spanish in the same conflict a few years prior to Fawkes. While the former was motivated primarily by material gain and career-related considerations, the latter was a soldier of conscience. The late-sixteenth- and early-seventeenth-century English state could do very little to prevent either type of foreign military enlistment.[55]

Toward the end of the seventeenth century, the English state allowed an army of Irish Catholics to leave Ireland and go to France. The Treaty of Limerick (1691) enabled between 16,000 and 19,000 soldiers (and their families), who had fought unsuccessfully for the restoration of the recently deposed Catholic king of England, James II, to join forces with King Louis XIV of France. Known in Ireland as the "Wild Geese," these soldiers lent credibility to attempts by the exiled pretenders of the House of Stuart to recapture the British throne—attempts that continued until 1745–1746. More importantly for our purpose, the Wild Geese served in the armies of France, Spain, and Austria.[56]

Initially motivated by loyalty to religion and monarch, the migration of Catholic soldiers from Ireland to France and elsewhere continued into the eighteenth century. In some cases in the 1710s and 1720s, recruits were assured that they would be able to serve the "Jacobite" cause (the attempts by the Catholic descendants of James II to recapture the British throne). However, motivations for traveling to the continent and enlisting were varied. Clandestine recruitment networks that depended on contacts in a number of locales meant that there was a strong family and regional basis for enlistment into the Irish regiments in Europe. The relative poverty of Ireland and the fact that Irish Catholics were all but barred from holding a commission in the British army, compelled many young men to seek a military career abroad. On average, about 1,000 recruits per year left Ireland between 1700 and 1730. According to Thomas Bartlett and Keith Jeffery, "the proper weight to be accorded the prospect of adventure (and plunder), the urge to escape a humdrum existence, even the simple desire to emigrate, cannot now be determined but should not be ignored."[57] Individuals embarking on foreign military service—in both the early-modern and modern periods—were often motivated by both the pull of ideology and adventure and the push of difficult or frustrating circumstances at home.

Alongside the gradual rise of nationalism, the eighteenth century also saw states beginning to take more resolute steps to restrict the recruitment of their subjects into foreign armies. In 1722 the British-dominated Irish Parliament banned unlicensed recruiting for foreign service. A number of recruiters were tried over the following years, and a large handful were sentenced to death and executed.[58] In 1736 King George II issued an act to put a stop to men being "seduced to enlist themselves to serve as soldiers under foreign Princes," without first obtaining license to do so. In 1746 Irishmen serving in the French or Spanish armies were barred from owning property in Ireland. Finally, in 1756, against the backdrop of the outbreak of the Seven Years' War, a new act explicitly forbade British subjects from enlisting in the military service of the French king either as officers or as noncommissioned officers.[59]

These legal restrictions, alongside improvements in the local economy and the opening up of the British army to Catholic recruits, led to a sharp decline in the number of Irishmen who joined the French army in the second half of the eighteenth century. Furthermore, the "Jacobite" cause that had attracted Irishmen to France was no longer a viable one. By the outbreak of the French Revolution, only the officers of the Irish regiments were predominantly Irish, and even these often came not from Ireland but from Irish communities that had settled in France. The rank and file were drawn from across western Europe. As we have seen, the distinctive identity of the Irish regiments in the French army was abolished in 1791, following the Revolution.[60]

As these examples from the First Crusade to the Wild Geese illustrate, ideals—in these cases, religious ones—played a part in the decision of individuals and groups that traveled abroad to fight long before the late eighteenth century. Yet, considering the total number of combatants in the wars of the Middle Ages and the early-modern period, the percentage of people who fought neither for material gain nor for their own ruler but primarily for the sake of ideals was very small. The percentage of foreign volunteers would also remain very small in the wars of the nineteenth, twentieth, and early twenty-first centuries. This raises the question: how do the earlier examples differ from the modern ones? The answer lies in the changing landscape of cultural norms and political expectations.

The development of the expectation that citizens should serve only in the armed forces of their own country, and not in any other, depended on the emergence of two interrelated assumptions. First, the ideology of nationalism, with its belief that each member of the nation has a stake in the collective's future, had to evolve. Second, the state had to assume enough power to demand a monopoly over the military service of its citizens. By the late eighteenth century, such expectations began to take hold, with varying degrees of pervasiveness, in western Europe, the United States, and elsewhere. Over the next two centuries these expectations became the global norm, even though they were never fully realized.[61] The rest of this book tells the story of foreign volunteers who continued to defy the norms and expectations of nation-states.

The Militarization of Liberty

Foreign volunteers have been a facet of modern warfare ever since the late eighteenth-century armed struggles that are normally associated with the birth of "national" armies. Their personal motivations for enlisting were diverse, but the principle of liberty was a recurring theme in their considerations. Liberty, understood as the freedom from arbitrary, despotic, or tyrannical government, was part of the value system of European aristocrats in the late eighteenth century. Aristocrats, or at least some of them, saw it as their duty to oppose a king who aspired to do away with the liberties of the realm. The same concept of liberty animated the leadership of the American Revolution. Indeed, the latter fought against "absolute Tyranny" and "absolute Despotism," to use the terminology of the Declaration of Independence.[62]

Following the outbreak of the revolution, hundreds of Europeans offered their military services to emissaries of the North American colonies in Paris. Benjamin Franklin, who served as the diplomatic representative to the court of Versailles between late 1776 and 1785, received 415 applications. Most of the applicants were French officers who sought a commission from the Continental Congress. While Franklin rejected most of these applications, and even though in March 1777 Congress adopted a resolution discouraging the acceptance of foreign volunteers, several Europeans did serve in the American War of Independence.[63] Two well-known foreigners who gained international fame for volunteering to fight for liberty in this way were the Pole Andrew Thaddeus Bonaventure Kosciuszko and the Frenchman Gilbert du Motier, Marquis de Lafayette.

Kosciuszko was born in 1746 to a family of minor gentry. His first step toward a military career was entering the Corps of Cadets at Warsaw. In 1769 he traveled to France, where he studied engineering and artillery but also attended the Royal Academy of Painting and Sculpture. He returned to Poland in 1774 only to find that his military skills were not in demand. In 1776, when news of the conflict between Britain and its American colonies began to filter through to Europe, the unemployed Kosciuszko was back in Paris. Silas Deane, an agent of the American revolutionaries in France who procured weapons and recruited volunteers, recommended Kosciuszko to Congress. The 30-year-old Polish officer arrived in Philadelphia in August 1776. Still relatively unknown and often mistaken for being a French officer, he managed to obtain a commission as colonel in the Engineers at sixty dollars per month.[64]

General George Washington was unenthusiastic about the foreigners who offered to join his forces, dubbing them "military fortune-hunters," "adventurers," "men of great ambition," or "mere spies." He complained of

> men who, in the first instance, tell you they wish for nothing more than the honor of serving in so glorious a cause as volunteers, the next day solicit rank without pay, the day following want money advanced to them, and in the course of a week want further promotion, and are not satisfied with any thing you can do for them.[65]

Kosciuszko, therefore, had to work hard to prove himself (an experience many other foreign volunteers came to share). As a military engineer, he worked on defenses and fortifications at Philadelphia, Ticonderoga, and Bemis Heights in the Saratoga, New York, area. Next, he was employed for more than two years in the planning and construction of fortifications at West

Point. As early as 1779 he referred to himself in a letter as "more than half a Yankee."[66] By the end of the war he enjoyed widespread acclaim and was promoted to the rank of brigadier-general. He was later awarded $18,940.25 in back pay and allowances and granted 500 acres of Ohio land as a veteran. Back in Europe, Kosciuszko tried to use his prestige to restore to Poland the lands that had been annexed by its neighbors. He led a failed uprising against Russia and Prussia in 1794, after which Poland was partitioned and lost what remained of its independence. Kosciuszko lived the remainder of his life in exile, continuing to promote the cause of independent Poland. He first tried to enlist the support of Napoleon, and later, at the Congress of Vienna (1815), petitioned Tsar Alexander I of Russia, but to no avail. He died in Switzerland in 1817.[67]

Eleven years younger than Kosciuszko, Lafayette came from the heart of ancien régime France. His family had a long military tradition, and he was commissioned as an officer at the tender age of thirteen. He was a young man of great ambition. "You ask me when I first longed for glory and liberty; I can recall no time in my life when I did not love stories of glorious deeds, or have dreams of traveling the world in search of fame," he wrote in an early draft of his memoir. When he first learned of the conflict in North America his "heart was enlisted": "Never before had such a glorious cause attracted the attention of mankind; it was the final struggle of liberty."[68] On 7 December 1776 he signed an agreement with Silas Deane in Paris that recorded his "wish to distinguish himself in this war and to render himself as useful as he possibly can." Noting "his Zeal for the Liberty of our Provinces," he was promised the rank of major general, even though he was only 19 years old.[69] The journey to the seat of war proved more difficult than expected. Suffering from financial difficulties owing to a series of military defeats, the American envoys in Paris could not afford to dispatch a ship carrying volunteers as they had hoped. Lafayette offered to find money to buy and arm a ship. His family objected to his plan and even convinced the French king to explicitly forbid him to leave for America. Lafayette left nonetheless.

His initial reception at Philadelphia was cold. "The Americans were disgusted by the conduct of several Frenchmen, and revolted by their pretensions," he noted. He appealed to Congress: "After the sacrifices I have made, I have the right to exact two favors: one is to serve at my own expense, and the other is to begin to serve as a volunteer."[70] His "zeal, illustrious family

and connexions" convinced his hosts to grant him the rank he had been promised on 31 July 1777.[71] Lafayette went on to form a cordial relationship with
General Washington. He distinguished himself on the battlefield, commanded American troops, enjoyed tremendous popularity, and played an
instrumental role in bringing France into the conflict against Britain in 1778.
Among the many honors and gifts he received after the American victory
was citizenship of the United States, a concept that did not yet exist in his
native France. Lafayette continued to exhibit his passion for liberty. Indeed,
both Kosciuszko and Lafayette took a stance against slavery. Lafayette went
on to play a pivotal role during the early stages of the French Revolution,
leading the newly formed National Guard. However, he objected to the abo-
lition of the monarchy in 1792 and eventually had to flee from France. The
revolutionary leaders who emerged in the 1790s saw liberty in more radical
terms.

One of the paradoxes of the French Revolution is that it not only nation-
alized military service but also militarized the idea of liberty. A key figure
in the militarization of liberty in revolutionary France was Jacques Pierre
Brissot de Warville. Brissot had visited the United States in the 1780s and was
impressed by what he saw. A leading member of the Legislative Assembly
and of the National Convention that replaced it in September 1792, Brissot
wrote passionately on the positive effects war could have for France and for
the revolutionary cause. As early as July 1791 he called on his compatriots to
follow the example set by the American revolutionaries:

> What soldiers of despotism . . . can for any length of time with-
> stand the soldiers of liberty! The soldiers of tyranny are after pay,
> they have little fidelity, and desert on the first occasion. The soldier
> of liberty fears neither fatigue, danger, nor hunger—he runs, he flies
> at the cry of liberty.[72]

Soon a pro-war faction gathered around Brissot while a rival anti-war party
also emerged, led by Maximilien Robespierre. In December 1791 Brissot's
paper, *Patriote français*, clamored:

> War! War! Such is the cry of all French patriots . . . such is the de-
> sire of all the friends of liberty scattered all over Europe, who are
> only waiting that happy diversion in order to attack and overthrow

their tyrants. It is that expiatory war which is to renew the face of
the world and plant the standard of liberty upon the palaces of kings,
upon the seraglios of tyrants . . . and upon the temples of popes and
muftis.[73]

Concerning the use of foreigners in the army, Brissot supported the abolish-
ment of the old foreign regiments, which he suspected of remaining loyal to
the ancien régime. For instance, he described the Swiss Guards as "an isolated and particular force, foreign to our principles, to our system of government."[74] However, as part of the "crusade for universal liberty" that Brissot
hoped to launch, he gave his assent to the formation of foreign legions, composed initially of Liégeois, Belgian, and Dutch exiled "patriots" who sought
to fight for the revolution. The Liégeois and Belgian legions—which began
to establish themselves even before receiving approval from the Legislative
Assembly in spring 1792—were composed of 1,150 men in all. The Dutch or
"Batavian" legions numbered 2,822. A further 2,160 men from Savoy and Piedmont were recruited in August 1792. A German brigade (later renamed the
German Legion) was created during the same month and included approximately 2,500 recruits. From Brissot's perspective, the enlistment of foreign
volunteers made for good propaganda. Their adherence to revolutionary
principles could be used to illustrate that the French were fighting for a universal cause. Furthermore, the legions were seen as a potential spearhead in
the attempt to "liberate" France's neighbors. Finally, there was also a practical consideration, which the revolutionary leadership shared with the military establishment of the ancien régime. Recruitment abroad strengthened
French forces at the expense of foreign powers.[75]

In creating the foreign legions, Brissot received the enthusiastic support
of General Charles-François Dumouriez. France, as Brissot explained to Dumouriez, was engaged in "a battle to the death between liberty and tyranny."[76]
Ambitious and opportunistic, Dumouriez served briefly as foreign minister
in the spring of 1792 before assuming command of the army in the north of
France in the summer of that year. Having captured Brussels, Dumouriez
was planning to invade the United Provinces of the Netherlands. He was also
the driving force behind the formation of the foreign legions, providing
the necessary political support as well as ministerial funds while these
were still at his disposal. Indeed, with Dumouriez heading his own army,

the revolutionary leaders in Paris, including Brissot himself, soon began to worry about the general's true intentions. They feared that he was seeking to create his own fiefdom in the Low Countries, probably with good cause. When Dumouriez ordered the creation of twenty-five new Belgian battalions, each consisting of 800 men, the Convention refused to relinquish all the necessary funds. In mid-December 1792, the Convention curtailed the autonomy of the army's generals by ordering them to apply the legislation of the revolutionary assembly in any territories they captured.[77]

The fall of Dumouriez came soon afterward, and the project of recruiting foreign legions, which he had championed, was soon scrapped. Entering the Netherlands in February 1793, Dumouriez was defeated by an Austrian army in March. Instead of facing the music in Paris, the general crossed over to the Austrian side, spending the rest of his life in exile. Back in France, the faction that had supported the war, Brissot and the Girondists, suffered the consequences. Brissot and many of his associates were arrested and executed. Robespierre and his Jacobin supporters were suspicious of foreigners from the start and did not believe that these could be trusted to defend the *patrie*. The defection of Dumouriez cast a new shadow of doubt over the reliability of the legions that were associated with him. The dismantling of the foreign legions began in autumn 1793, under the "Reign of Terror." The depleted Belgian and Dutch units, which in any case struggled to recruit enough men of the right nationality (incorporating French, Prussian, Austrian, and even British soldiers instead), were disbanded. The men were reorganized into units identified only by numbers, thus making no reference to their foreign origin. Finally, Article 287 of the Constitution of 1795 forbade the recruitment of foreign troops.[78]

But focusing on the French side only reveals half the story. To understand the personalities and motivations of foreign volunteers who fought for the Revolution, a few individual stories must be considered. One of the best known foreigners who took up arms for revolutionary France was Sebastián Francisco de Miranda. According to his biographer, he was an opportunistic man who was nonetheless "passionately devoted to the ideas of liberty and freedom."[79] Miranda was born in 1750 to Creole parents in Caracas, Venezuela, then still a Spanish dominion. In 1772 he moved to Spain and purchased a commission as a captain in the Spanish infantry for 8,000 pesos (the practice of officers purchasing their commissions was still prevalent in eighteenth-century European armies). In 1774–1775 he fought with Spanish forces in North Africa, and in 1780 he was dispatched to the West Indies and North America. Spain was allied with France, so Miranda fought against the British in Pensacola, Florida. While fighting alongside the American revolutionaries, he became animated with the idea of liberating South America from its Spanish rulers. Quarrelsome by nature, Miranda fell out of favor with his superiors. He was arrested in Cuba in 1782 and deprived of his commission in 1783. He subsequently decided to tour the United States. Passing through New Hampshire in October 1784, he noted that even though the local population had to contend with poor pastures, "the spirit of Liberty inspires these people," making them happier than the rich owners of mines and land in Mexico, Peru, Buenos Aires, Caracas "and the whole of the Spanish-American continent."[80] After his North American tour he traveled extensively in Europe. Miranda was rumored to have had an affair with Catherine the Great during his visit to Russia in 1787.[81]

When Miranda arrived in revolutionary Paris, he befriended Brissot. By August 1792 Miranda had entered into French military service as a lieutenant-general, having agreed to "serve the cause of liberty."[82] Dumouriez initially took a liking to the "Peruvian" who became his second in command in the Army of the North. However, when Dumouriez sat down to write his memoir from his exile in London, his appraisal of Miranda was not very flattering:

> He was better versed in the theory of war than any other of the
> French generals but he was not equally well versed in the practice . . .
> he was unfit to command the French, whose confidence it is impossible to gain but by good humour and a conduct expressive of respect
> for them.[83]

Dumouriez held Miranda partially responsible for his defeat in the Netherlands in 1793. But while Dumouriez fled to exile, Miranda was arrested and was unable to leave France until 1798. He then settled in Britain and campaigned for British support for the liberation of Spanish America. In 1810 he returned to Caracas. Elected to the national congress, he was present when the first republic of Venezuela was officially declared on 5 July 1811. Following a devastating earthquake in Caracas, Miranda was named generalissimo and head of state with sweeping powers over the army in April 1812. After briefly commanding the pro-independence republican forces, which fought against

both the Spanish and local royalists, Miranda was captured (one of the officers who handed him over to the Spanish was the future South American
leader Simón Bolívar). He died in prison outside Cadiz in 1816.[84]

A less well-known foreign volunteer who joined the French army following the revolution was the Scotsman John Oswald. Born in Edinburgh
in 1760, he enlisted in the British army in 1776 or 1777. He soon obtained the
rank of lieutenant, probably after purchasing his commission. In 1781 his regiment was sent to India where he not only saw action but was also impressed
by local customs and traditions. He adopted vegetarianism and later wrote
a treatise in its praise. Like Miranda, he traveled extensively before returning
to Britain, where he wrote for various periodicals. His political views were
very radical by the standards of the day. He opposed the hereditary accumu-
lation of land by private proprietors and espoused active and direct, rather
than representative, democracy.[85]

A staunch supporter of the French Revolution—first in London and subsequently in Paris—Oswald befriended Brissot, with whom he shared the belief in universal liberty. Indeed, in 1793 Oswald published a suggestion for a
constitution for a universal republic. He also held firm views about how the
Revolution ought to defend itself. After war broke out in 1792 he wrote a military treatise that highlighted the merits of a popular militia over standing
professional armies. The title of the work is very telling: *Tactics for the People,
or, New Principles for Military Evolutions, which will Teach the Masses how to Learn
to Fight by Themselves and for Themselves, without the Dangerous Use of Standing
Armies.* The rationale for this form of military organization drew from his
concept of direct democracy. Just as people should not relinquish their political decision making to representatives who would "think instead of them,"
they also should not leave their defense in the hands of a standing army that
might turn against them.[86] Oswald practiced what he preached, devoting his
military experience to and eventually sacrificing his life for the Revolution.
He raised a battalion and trained it in the use of pikes. In 1793 he died in battle
while trying to put down the large-scale revolt that had broken out in La
Vendée in western France. The use of foreign troops in the suppression of
domestic revolts was yet another way in which the revolutionary leadership
replicated the military practices of the ancien régime.[87]

The experiment of recruiting foreign volunteers into the French army
after the Revolution highlighted a number of facets that are typical of the

phenomenon more broadly. First, even ostensibly "national" wars could attract foreign recruits, especially when broader transnational ideals such as liberty seemed to be at stake. Second, there was a risk that the enthusiasm of foreign volunteers could be harnessed for cynical, self-serving purposes, as they had been under General Dumouriez. A third aspect worth highlighting is how the willingness of host states to countenance foreign units was subject to changes in the political climate. What seemed acceptable and perhaps even opportune in the spring of 1792 quickly went out of fashion a year later. As the Reign of Terror subsided and the era of Napoleon was about to begin, foreign volunteers for liberty were given a second chance under French arms.

Exiled Polish generals, stateless after the partition of their country, submitted their plans for the creation of a Polish legion to the directory that governed France in November 1795. Such a legion, they argued, would deprive France's enemies of recruits, form the nucleus of a Polish republican army, and help spread the ideals of the French Revolution. However, a reluctant French leadership was able to quote Article 287 of the Constitution of 1795 to explain its refusal to create such a foreign legion. A year later General Jan Henryk Dabrowski and the poet Jozef Wybicki were able to secure the directory's approval for the formation of a distinctly Polish force. By early 1797 a first legion was raised, drawing on thousands of Polish émigrés and former prisoners of war (the latter were captured by the French, having served in the Austrian army). The legion's uniforms bore the inscription "free men are brothers."[88] It fought for the French in Italy through to 1799 and suffered immense losses. Their exploits went on to inspire generations of Poles, and Wybicki's song that they marched to, *Poland Has Not Perished While We Live*, became the Polish national anthem in the twentieth century. In 1799 a second Polish legion was raised, with the French justifying their move by arguing, "if the coalesced kings deploy vast armies against free peoples, the latter must admit into their ranks all men whom a sublime fervour calls to fight for the sacred cause of liberty."[89]

The rise to power of Napoleon, who had been instrumental in setting up the first Polish legion, initially gave the exiled Poles hope. Dabrowski tried to convince Napoleon to allow the legionaries, whose numbers swelled to 13,000, to fight for the freedom of Poland. However, such a scheme did not fit into Napoleon's strategic plan. In October 1801 the latter concluded a treaty with Alexander I of Russia. Article 3 of this treaty bound both parties to prevent precisely the sort of activities the exiled Polish leaders called for. In 1802–1803 many of the disgruntled legionaries were sent from Italy to Haiti as part of Napoleon's attempt to put down the Haitian Revolution. For fear of mutiny, they were not told of their destination until they were out at sea. Once in Haiti, it soon transpired that the Polish troops were ill-prepared for the climate and for guerrilla warfare. Suffering from low morale, one French officer commented: "The Poles are abominably bad soldiers for the kind of war we are waging here."[90] Casualties in combat and from yellow fever as well as desertion reduced the force of over 5,000 to a few hundred. By this stage the aging Kosciuszko, living in exile in France, became completely disillusioned with the idea that Napoleon would restore Poland: "he thinks only of himself. He hates every great nationality and still more the spirit of independence."[91]

The fate of the Polish legions under Napoleon again illustrates how the willingness of foreign volunteers to fight could be manipulated and used for self-serving purposes. Napoleon also created an Irish Legion in 1803, ostensibly to be used in the liberation of Ireland from British rule. In practice, the few Irishmen who enlisted were deployed in Russia and the Iberian Peninsula.[92] Regarding the motivations of the volunteers, we can see here how, not for the last time, individuals and groups were prepared to fight in one war (as the Poles did in Italy) in order to have a chance to fight for their homeland later on. The ideal of liberty, even though it was understood in different ways by different people, continued to inspire foreign volunteers throughout the nineteenth century.

* 2 *

Attractive Conflicts

The Changing Ideological Landscape

Vicarious Plight

In 1937 a brutal civil war was raging in Spain. The conflict elicited strong emotive responses hundreds and even thousands of miles away.

> What is the price the gallant Spanish people is paying for its stirring defense of Democracy? Thousands have died . . . . For them we can do little, save honor their memory. Tens of thousands—soldiers and civilians—have been wounded. The cruel terror that sweeps out of the skies has left women and children stricken on the pavements of beleaguered cities. Their suffering can be mitigated, their healing speeded.[1]

These words, taken from a pamphlet by the New York–based Medical Bureau of the American Friends of Spanish Democracy, were part of a campaign to raise money for the creation of an American "base hospital" in Spain. Prior to this, the Medical Bureau had already dispatched ambulances, supplies, and medical staff to Spain, to assist the Republican cause. One of the additional objectives of this and various other fund-raising campaigns was to arouse public opinion sufficiently to end the boycott of military aid to Republican Spain imposed by the United States government.[2]

Foreign war volunteers often respond to media coverage of the conflict they eventually travel to join. Other sympathizers with the cause in question might respond by taking part in fund-raising activities or petitions and demonstrations calling on their government to take firmer action in support of one side in a war between states or within another state. Volunteering to fight abroad is only one, perhaps extreme, manifestation of solidarity. The link between activism at home and volunteering to fight abroad was captured by Fred Thomas from northeast London, who joined the International Brigades in the spring of 1937. "The situation in Spain called for every one of us who opposed fascism to do his utmost to defeat it," he wrote in his memoir. An active Labour Party member, Thomas had been taking part in antifascist gatherings even before the war in Spain broke out. However, once it did, he increasingly felt that "meetings, demonstrations, collecting food, money, medical supplies, seemed no longer the answer. I determined to go to Spain myself."[3] Volunteers like Thomas also had personal, non-ideological reasons for choosing such a risky path. These are discussed in Chapter 3. For now let us focus on the causes that have impelled foreign volunteers into action.

Foreign volunteers nearly always mobilize to defend something: another country's independence or democratic form of government; fellow socialists or fellow fascists; members of an ethnic, national, or religious group; and so on. Some sort of vicarious plight appears, at least initially, to be a crucial component in bringing people from other countries to risk their lives in foreign lands. Of course, plight resonates differently depending on the outlook and sympathies of the receiving audience. And audiences invariably exercise a selective and biased understanding of world affairs. For instance, the plight of Republican Spain in 1936, under attack from a military coup that was supported by monarchists, Catholics, traditionalists, and fascists (both in Spain and abroad), resonated with many socialists, communists, anarchists, and other antifascists in various parts of the world. In contrast, the persecution of the Church behind Republican lines, which included the murders of thousands of priests and clerics as well as the burning of religious buildings, enraged many Catholics outside Spain.[4] The Spanish Civil War was not unique in splitting world opinion. When violent clashes began between Arabs and Jews in Palestine in late 1947 and early 1948, many Jews in North and South America, Europe, and South Africa rallied in support of their co-ethnics.

Meanwhile public opinion in the predominantly Arab cities of the Middle East was decidedly on the side of the Palestinian Arabs.[5]

Some wars were more successful than others at attracting foreign volunteers. This "success" was not necessarily determined by the amount of suffering each war produced, but by the availability of audiences abroad that were sympathetic toward the plight of groups participating in the conflict. A glance at the wars that were fought in the nineteenth, twentieth, and twenty-first centuries reveals that the conflicts that attracted substantial numbers of foreign volunteers normally involved a salient ideological fault line. Such "attractive" conflicts brought to the fore ideological confrontations that had an appeal and a relevance that transcended state boundaries. These conflicts can be divided, by way of generalization, into three waves. The political scientist David C. Rapoport, who used the concept of waves to depict different stages in the history of international terrorism, described the term as a "cycle of activity in a given time period." During each cycle, "similar activities occur in different countries, driven by a common predominant energy that shapes the participating groups' characteristics."[6]

Much like Rapoport's waves, the waves of ideological fault-line conflicts that attracted foreign volunteers were consecutive and overlapping. The first wave, which continued a trend established in the late eighteenth century, pitted "liberty" against "tyranny." In the decades following the defeat of Napoleon and the Congress of Vienna (1815), many foreign volunteers came from the ranks of those who sympathized with the plight, and adopted the cause, of national movements struggling against monarchies and imperial powers that were perceived as stifling liberty. Conflicts in the first wave include the wars of independence in Spanish America, the successive wars between the Ottoman Empire and the Greeks or other peoples of the Balkans, and the wars of Italian unification. The second wave, which began following the October Revolution in Russia (1917), pitted the Left against the Right. Foreign volunteers from both sides of the political spectrum fought either to assist or to prevent regime change in another country or group of countries. Conflicts in this wave include the Russian Civil War, the Polish-Soviet War, the Spanish Civil War, the Second World War, and the protracted Cold War conflicts between right-wing and left-wing forces in Africa, Asia, and Latin America. The third wave could be bound together under the banner of a clash of civilizations. Starting in the 1980s with the Soviet-Afghan

War, this wave includes the Yugoslav Wars, the conflict in Chechnya, the post-2001 war in Afghanistan, the post-2003 war in Iraq, and the still ongoing (at the time of writing) civil war in Syria.

There are, of course, exceptions. For instance, the first Arab-Israeli War in 1948, which according to this periodization took place during the second wave, could hardly be depicted as a struggle between Left and Right. Initially a civil war between Palestinian Arabs and Jews, it became a war between states after the Israeli declaration of independence and the subsequent invasion by the neighboring Arab states in mid-May of that year. There are also lines of continuity that run through the different waves. The conflict in eastern Ukraine that erupted in 2014, for example, incorporated elements from the second wave, with some foreign volunteers defining themselves in terms of leftist and rightist political allegiances. Ironically, the pro-Russian Donbass rebels, who are fighting against forces loyal to the Ukrainian government, have hosted far-right French volunteers, who saw Russia as a bulwark against liberal globalization, as well as leftist volunteers from Spain, one of whom showed off tattoos of the Soviet leaders Vladimir Ilyich Lenin and Joseph Stalin on his torso.[7] As this recent example reminds us, armed conflicts create realities that can be far more complex than the straightforward ideological clarity with which foreign volunteers and their supporters perceive them. Be that as it may, dominant ideals have had an undeniable influence on the decision of foreigners to volunteer in each of the three waves.

Liberty versus Tyranny

In 1814 the Bourbon dynasty was restored in Spain. King Ferdinand VII sought to curb many of the liberties that had sprung up during the Napoleonic Wars. He rescinded the constitution of 1812, dissolved the national assembly—the Cortes—at Cadiz, and reinstated the inquisition. This approach was received with indignation by political elites in Spanish American territories, provoking a series of independence proclamations. Spanish American independence documents were liberal and democratic in character, inspired to a large extent by the United States and the French Revolution.[8]

Agents representing the newly proclaimed states approached the United States and Britain, asking for political and material support in their struggle against Spain. Because of their international obligations toward Spain, the governments of both Britain and the United States declared their neutrality in the conflicts that ensued. However, in both countries there was broad support for Spanish American independence. For instance, in 1821 the Speaker of the House of Representatives, Henry Clay, argued that the United States had a moral duty to "give additional tone, and hope, and confidence to the friends of liberty throughout the world." He believed his country ought to support Spanish American independence by all means possible, short of direct military intervention.[9] Similarly, in Britain, parliamentarian Sir James Mackintosh cited a petition signed by 107 businessmen from London who wanted to trade with the new states of South America in an attempt to convince the government formally to recognize their independence. "Are we, the English nation, to look thus coldly on rising liberty?" he asked during a debate in the House of Commons in 1824.[10]

The South American leader Simón Bolívar asked his representative in London to raise troops for the liberation of Venezuela in April 1817. This request proved easy to fulfill. Almost 7,000 left Europe between 1816 and 1822 to join the army of what became known as Gran Colombia (roughly corresponding to modern-day Colombia, Ecuador, Panama, and Venezuela). A further 1,000 joined Bolívar's naval forces. A belief in liberty inspired some to volunteer while others were more interested in pay, plunder, adventure, or the promise of land. Indeed, emigration, for all manner of personal, social, or economic reasons, was an integral part in the decision of many to enlist. A mixture of idealism and material incentive was evident in the terminology used by the individuals who recruited volunteers for Bolívar's cause. For instance, James Towers English, who had already fought in South America and returned to Britain to raise men for a British Legion in 1819, declared to his recruits:

> Remember you are Britons, Remember the Eyes of your Country are fixed upon you, and wish for your Success. . . . You are already Hali'd by those brave and suffering People [as] the Liberators of their long oppressed Country. They are prepared to open their arms to receive you: all that Tyrants and Oppression have left them (Their Country), the Richest and most fertile in the world, they offer to share with you.[11]

Once in South America, the maintenance of these foreign recruits proved costly and difficult. Bolívar's victories in Gran Colombia enabled him to cease foreign recruitment in the early 1820s.[12]

The struggle for independence in the "Southern Cone" also attracted foreign volunteers, though these were fewer than in Bolívar's campaigns in the northern part of the continent. A few British volunteers served in prominent positions in General José de San Martín's Army of the Andes, which was raised in the United Provinces of the Río de la Plata (nowadays Argentina) and used to liberate Chile from the Spanish Empire.[13] But probably the most famous foreign volunteer who became involved in the wars of independence in South America was the British naval officer Lord Thomas Cochrane. Cochrane first earned his fame during the Napoleonic Wars. His deeds led his French enemies to nickname him *le loup des mers* (the wolf of the seas). Radical in his political leanings but also not averse to amassing personal fortune, Cochrane was implicated in a stock exchange fraud in 1814 and was dismissed from the Royal Navy. In 1818 he accepted the offer of the Chilean leader Bernardo O'Higgins to organize and head a Chilean navy. He played a key role in Chile's war against Spain and also contributed to the independence of Peru, Spain's last stronghold in South America. When Cochrane took over, roughly 15 percent of the Chilean navy was composed of foreigners, with their share rising to about one-third by the time he left.[14] Following the demise of Spanish power in South America, Brazil rose in revolt against Portugal. In the autumn of 1822 Cochrane received an invitation to command the Brazilian navy. The Brazilian agent in Buenos Aires implored him to "help to tame the Hundred-Headed Hydra of frightful Despotism."[15] Cochrane accepted. His vessels were able to blockade Portuguese garrisons along the Atlantic coast and to force their surrender.

Cochrane and a few notable exceptions aside, the foreigners who fought in the South American wars of independence were often associated in the 1820s with desertion, mutiny, mercenarism, and adventurism. More than half of them either died of yellow fever and poor provisions, returned home without firing a shot, or deserted to seek their fortune in the Caribbean.[16] At the same time a new cause had emerged that attracted the attention of individuals who wished to support the struggle, and in some cases fight, for liberty: the Greek revolt against the Ottoman Empire.

In March 1821 Pietro Mauromichali, a Greek commander in the Peloponnese, issued a manifesto to the governments and peoples of Europe:

> Reduced to a condition so pitiable, deprived of every right, we have, with unanimous voice, resolved to take up arms, and struggle against the tyrants. . . . In one word, we are unanimously resolved on Liberty or Death. Thus determined, we earnestly invite the united aid of all civilized nations to promote the attainment of our holy and legitimate purpose, the recovery of our rights, and the revival of our unhappy nation.[17]

The Greek revolutionaries' call for assistance, often echoed and magnified by Greek diaspora communities across Europe, struck a chord with individuals and groups coming from a variety of backgrounds. Ancient Greece had a powerful symbolic value for upper-class Europeans who had been raised in the classic tradition. Homer, ancient Athens, Sparta, Pericles, Sophocles, and Plato were part and parcel of educated Europeans' cultural frame of reference. Hence, many desired to support the struggle of the modern Greeks, who were seen as the descendants of the ancient fathers of Western civilization.[18]

Greece was also heavily romanticized. From the eighteenth century onward, many western European travelers, fortunate enough to embark on a Grand Tour, aspired to visit Greece. One notable visitor in 1809 and 1810 was the young British poet Lord George Gordon Byron. Byron's poem *Childe Harold's Pilgrimage*, which was based on his travels and published in 1812, became a bestseller, was translated into several languages, and turned its author into a European celebrity. The Greece that emerges from the poem lies in ruins:

> Ancient in days! august Athena! where,
> Where are thy men of might, thy grand in soul?
> Gone—glimmering through the dream of things that were . . . [19]

But it was precisely these ruins that lent Greece its poetic value. As Byron later explained in a letter to his publisher, "it is the '*art*,' the columns, the temples, the wrecked vessel, which give them [the Acropolis and other ancient sites] their antique and their modern poetry."[20] The aspiration that

Greece might regain its freedom appeared in a number of Byron's poems, including *Don Juan*.

There was also a religious element in European solidarity toward the Greeks in their struggle against the Ottomans. The coverage of the conflict emphasized the cruelty and repression enacted by the Muslim Turks and often overlooked violent atrocities carried out by the Christian-Orthodox Greeks. The hanging of the Greek Orthodox patriarch of Constantinople at the gate of his own palace in April 1821, despite his denunciation of the revolt, was one of a number of incidents that were portrayed very negatively in the western European press.[21] For instance, on 1 June the *Times* of London, quoting reports from across Europe, argued that the revolt "carried the fanaticism and exasperation of the Turks to its height; and this people, greedy of blood, have abandoned themselves to the great excesses against defenseless Christians."[22]

Support for the Greek cause throughout the conflict manifested itself in different ways. On a humanitarian level, a number of committees were set up in Europe and in North America to raise funds and send supplies to Greece. At the same time a number of artists and poets addressed in their works the plight of the Greeks as well as atrocities committed against them. Ugo Foscolo, an Italian poet of Greek descent, wrote a canto, *For the Heroic Defence of Missolonghi*. The French artist Eugène Delacroix dedicated two oil paintings to the war in Greece: *Massacre at Chios* (1824) and *Greece on the Ruins of Missolonghi* (1826).[23] And then there were those who decided to take direct action.

The number of philhellenes, lovers of Greece who fought for the Greek cause, is estimated at between 1,000 and 1,200. The first cohort of foreigners who went to Greece did so at the beginning of the conflict, in 1821–1822. Later cohorts included the most famous volunteer of the Greek War of Independence, Lord Byron, and Lord Cochrane, the veteran of the recent wars in South America. After their arrival, the experiences of the philhellenes were often difficult; these are discussed later. For our present purpose, it is worthwhile to briefly examine who heeded the philhellenic call.

The foreign volunteers who fought for the independence of Greece were not a homogenous group. They included diaspora Greeks, some of whom went on to play leading political roles in the fledgling Greek state. Among the non-Greeks were French, Italian, German, and Polish unemployed or underemployed veterans of the Napoleonic Wars. There was also a large cohort of students and liberals from the German states as well as exiles from failed, anti-monarchical revolutions across Italy. As their diverse backgrounds suggest, for most of the volunteers, philhellenic sentiment was only one of the reasons that brought them to Greece.[24]

Perhaps the individual who best symbolized the nineteenth-century wave of foreign volunteers who fought for liberty was Giuseppe Garibaldi. Garibaldi led an extraordinary life that included stints as a revolutionary conspirator and a respected member of the Senate of the Kingdom of Italy, as a merchant seaman and a naval commander, as a guerrilla leader, and as an army general. He was born in Nice in 1807. After the fall of Napoleon, his city of birth, which had been annexed by the French, reverted to the Kingdom of Piedmont-Sardinia. While he was in his twenties Garibaldi joined the revolutionary movement Young Italy. This movement, led by the politician and intellectual Giuseppe Mazzini, aspired to bring about the unification of Italy under a republican regime. Mazzini's philosophy, according to which international peace could be secured by governments based on popular will and the natural goodness of ordinary people, resonated with the young and energetic Garibaldi. After a failed insurrection in Piedmont, Garibaldi was forced to flee in 1834, having been sentenced to death in absentia.[25]

He traveled to Brazil, where the southernmost province of Rio Grande do Sul had declared itself independent from the imperial government in Rio de Janeiro. Garibaldi served as a naval commander for the fledgling republic. In 1841, with the Empire of Brazil gaining the upper hand, Garibaldi moved to Uruguay. Uruguay soon became embroiled in a protracted conflict against its larger and stronger neighbor, the Argentine Confederation. While Uruguay boasted a democratic and civilian constitution, the Argentine leader General Juan Manuel de Rosas was considered by many foreign observers as a despot who did not shy away from employing terror to maintain his governorship. Garibaldi was instrumental in establishing a naval force, which attempted to disrupt Argentine shipping in the River Plate area, and he organized an Italian legion of about 500 or 600 volunteers to help in the military defense of Montevideo. Uruguayan forces also included a larger French legion as well as legions of Spanish and Basque volunteers, but the government appointed Garibaldi as commander-in-chief of its forces in 1846. This period of his life was not without its blemishes. He was forced to resign in July 1847, largely due to local resentment at having a foreigner at the head of the garrison. There were also misgivings about unauthorized requisitioning and the freebootery of his Montevidean flotilla. Garibaldi nonetheless emerged from the conflict as a hero for many Italian nationalists. His exploits in South America and the feats he went on to achieve in Europe earned him the accolade the "hero of two worlds."[26]

Garibaldi returned to Italy during the revolutionary year of 1848. In 1849 Pope Pius IX fled from Rome, and a short-lived Roman Republic was established. Garibaldi was appointed commander of the Republican troops, which fought against numerically superior French forces that tried to restore the pope to his temporal seat. By summer 1849 the republic had been crushed. Garibaldi was forced into exile once again. The next dramatic episode in his biography came in the late 1850s. Garibaldi broke with Mazzini, with whom he had collaborated in Rome. Setting aside the aspiration for republicanism in favor of achieving a more speedy unification of Italy, Garibaldi aligned with the Piedmontese monarchy and served as a general in the 1859 war against Austria. In 1860 he led an expedition of about 1,000 volunteers, dressed in red shirts that had become the Garibaldian symbol, to "liberate" southern Italy from its Spanish Bourbon rulers. The victories of his forces in Sicily contributed to his legend. For instance, one British newspaper commented that "history will preserve a bright and lasting page for the record of the heroism, bravery, daring, patriotism, and skill of the great general, Garibaldi."[27] Foreign volunteers from across Europe and North America flocked to join his advance toward Naples. At the height of his popularity Garibaldi surprised many by handing over the territories he had liberated to the newly established Kingdom of Italy and retiring to his farm on the island of Caprera.

The following decade was a restless one. The retired general tried, unsuccessfully, to lead volunteers to capture papal Rome, clashing with the forces of the Italian state (1862) and with papal troops (1867). He also remained highly active internationally. During the early stages of the American Civil War, Garibaldi corresponded with and was visited by Union officials who wanted him to join, and perhaps even lead, the forces of the North. He initially declared himself willing to "go to America, if I do not find myself occupied in the defense of my country."[28] However, he expected the Abraham Lincoln administration to emancipate the slaves long before the US president found it politically advisable to do so. Eventually Garibaldi decided not to travel to North America.[29] In Europe he lent his moral support to the Polish insurrection of 1863 and the Cretan revolt against the Ottoman Empire in 1866–1869. During the Franco-Prussian War of 1870–1871, after Napoleon III (whom he loathed) had been toppled and the French Third Republic was proclaimed, Garibaldi sprang into action once more. He volunteered to fight against the Prussians and commanded a force of local and foreign volunteers known as the Army of the Vosges in the Dijon area. As we shall see later on, the Garibaldi myth endured and continued to be evoked by generations of foreign volunteers, holding divergent political creeds, long after his death in 1882.

For now, it is worth highlighting the extent to which his contemporaries associated Garibaldi, who achieved international fame, with fighting oppression and upholding liberty and independence. Substantial amounts of money were collected in Britain and sent to Garibaldi between 1856 and the end of 1860. He had become a celebrated cause for British liberals, with both Florence Nightingale and Charles Dickens making donations. In September 1860 the British *Morning Post* expressed grave doubts about whether Francis II could hold on to his Kingdom of the Two Sicilies: "it is not feasible that Francis, Despotism, and Universal Mismanagement, can win the day against Garibaldi, Liberty and the prospect of a united Italy."[30] The French painter Ulric de Fonvielle, a staunch republican who was bitterly opposed to Napoleon III, went to Sicily to join the forces of Garibaldi. "I hear that liberty is awakening in a corner of the world, I run there . . . come what may!," he declared.[31]

But Garibaldi and the cause of liberty that he championed did not enjoy universal consensus. The way in which news from Italy was received depended on the political, social, and cultural outlook of the receiving audiences. For instance, conservative papers in Britain such as the *Newcastle Courant* were far more reserved and at times even critical in their coverage of Garibaldi's exploits.[32] And then there were those who were so alarmed by the prospect of a unified Italy absorbing Rome and annexing what remained of the Papal States that they decided to take direct action. Catholic volunteers from France, Britain (which included Ireland), the Netherlands, Belgium, Canada, Austria, the United States, and other countries rallied to defend the papacy. Several thousand foreign volunteers, who became known as Pontifical Zouaves, participated in the defense of Rome between 1860 and 1870, fighting against both the Italian state and Garibaldi's volunteers.[33]

The struggle for liberty acted as an umbrella cause in the conflicts in South America in the late 1810s, Greece in the 1820s, and throughout the Risorgimento (the process of Italian unification).[34] During the late nineteenth century this wave began to lose momentum. Of course, foreign volunteers fighting for liberty did not vanish overnight. Several hundred foreigners arrived in the Balkans in 1875–1876 following the outbreak of revolt in Herzegovina against the Ottoman Empire. Some of these volunteers stayed on to fight in the subsequent war between Serbia and Montenegro, on the one hand, and the Ottomans, on the other. A further 2,000 or so foreigners took part in the Greek war effort against the Ottomans in 1897.[35] However, the collapse of the Ottoman, Austro-Hungarian, and Russian Empires following the First World War and the 1917 October Revolution in Russia marked the end of the first, and gave birth to the second, wave of ideological fault-line conflicts that attracted foreign volunteers.

Left versus Right

In terms of the ideological appeal that armed conflicts exercised on audiences abroad, the coup d'état staged by the Bolshevik party in Russia in the autumn of 1917 was an important turning point. The communist regime that was established in what soon became known as the Soviet Union called for global revolution (though this policy was not always pursued very vigorously). In the seven decades that followed, how each side in an armed conflict positioned itself in relation to the idea of class struggle became a key indicator of the media coverage it received and the kind of transnational support it garnered.

Foreigners took part, either voluntarily or involuntarily, in the first ideological conflict that followed from the October Revolution—the Russian Civil War. In December 1917 Russia held approximately 2.1 million First World War prisoners from the multinational Austro-Hungarian Empire and a further 170,000 from Germany. While hundreds of thousands were repatriated in the months that followed, large numbers of former prisoners of war were mobilized by the different sides engaged in the Civil War. The newly established Red Army, which sought to protect the revolutionary regime, was in desperate need of experienced military manpower.[36] It recruited some former prisoners by force while others joined voluntarily, sensing an affinity with the Bolsheviks' political agenda of equality and land redistribution. The latter included a few individuals who later rose to political or literary prominence: the future leader of Yugoslavia, Josip Broz Tito; the Czech writer Jaroslav Hašek, author of *The Good Soldier Svejk;* and the Hungarian revolutionary Béla Kun.[37]

The rise of communism in Russia reverberated throughout Europe and beyond. In 1919 there were attempts to establish Bolshevik-like regimes in Bavaria and Hungary as well as strikes and armed clashes on a smaller scale elsewhere. In 1920 a Polish invasion into Ukraine was repulsed by the Red Army, with the latter launching a counteroffensive into Polish territory. Some Soviet leaders thought of Poland in terms of a "red bridge" over which the victorious revolutionary army was going to march out of Russia and into Europe.[38]

The specter of the advance of communism formed part of the motivations of American, Canadian, and French pilots and airmen who came to fight alongside the Poles against the Red Army. In his account of the history of Poland's foreign squadron, pilot Kenneth Malcolm Murray of New York spoke of "the brutality and lust of Russian Communism."[39] Although not all Poles saw the conflict with Russia in terms of a clash between communism and Christianity—the Polish leader Jozef Pilsudski had previously been a leftist revolutionary—Murray certainly did.[40] He argued that the Polish victory near Warsaw in the summer of 1920 "has since come to be known as the 'Eighteenth Decisive Battle of the World' for its historical significance to all Christians." Writing in 1932, he believed that "probably not for a very long time will it be possible to gather together again such a rare lot of *war-trained* pilots to fly under a single flag, nor for such a world-wide cause—the preservation of Christian civilization against the all-encompassing threat of Bolshevism."[41]

Despite Murray's prediction, the polarization of European politics in the 1930s ensured that foreign volunteers of all hues assembled only a few years after his book was published. In Italy, Benito Mussolini's Fascist regime, which had long presented itself as the antithesis to the "two red years" of civil unrest (1919–1920) in Italy, became much more militant following its 1935 invasion of Ethiopia. In Germany, Adolf Hitler and his Nazi Party, in power since 1933, persecuted ethnic minorities and leftist opponents, pursued a policy of military rearmament, and called for the revision of the borders and limitations imposed on their country by the Treaty of Versailles (1919). France was in political turmoil, with right-wing riots in 1934 that threatened to topple the Third Republic. In 1935 the Moscow-led Communist International (Comintern) encouraged communist parties abroad to form political alliances—or popular fronts—with moderate left-wing parties in an attempt to halt the advance of fascism. The conflict that brought the tensions between Right and Left into the open most vividly was the Spanish Civil War.

Harry Pollitt, the chairman of the Communist Party of Great Britain, exemplified this atmosphere of polarization in an article he published in late August 1936 under the title "To the Aid of the Spanish People!": "The whole future development of the international situation is being worked out in the struggle between the popular forces of democracy, law and order and the bestial, terroristic forces of fascism in Spain. . . . There can be no neutrality in this life and death struggle."[42] This rhetoric found an echo among the approximately 32,000 foreign volunteers who joined the International Brigades. Communist Party member Joseph Dallet Jr. from Ohio wrote to his mother in March 1937 to justify his decision to volunteer. He argued that Hitler and Mussolini's intervention in Spain "must receive a military defeat of the first order if the peace of Europe and the world is to be preserved and if fascism is to be checked, instead of spread."[43] Tom Wintringham, who commanded the Brigades' English Battalion in 1937, described his comrades as "part of a new fellowship growing in Europe and America and Asia: the fellowship of those who believe that our small and precarious beginnings of civilization, all we value of personal happiness and social freedom, are being destroyed by fascism."[44] For another British volunteer, Bill Alexander, the attachment to this cause was so strong that decades later he still insisted that "we did not go to Spain to fight in a 'Civil War'—we went to fight in an 'anti-fascist war.' "[45]

Meanwhile the enemies of the Spanish Republic and its Popular Front government tried hard to bolster their own claim for international legitimacy. On 27 July 1936 the right-wing British *Daily Mail* published a declaration by General Francisco Franco, who would soon emerge as the leader of the insurgents. Franco described the insurrection as a "nationalist movement to save Spain from Russian domination."[46] The insurgents presented themselves as the Nationalist side for both domestic and external reasons. By doing so they not only depicted their enemies as the henchmen of Joseph Stalin but were also able to draw on the memories of the war against Napoleon (1808–1814), where patriots and the Catholic Church fought side by side against a foreign invader. Furthermore, by using the term "Nationalist," they consciously tried to avoid their identification with fascism.[47]

The Nationalist message struck a chord with right-wing supporters and would-be volunteers. Irish general Eoin O'Duffy, founder of the anti-communist Blueshirts, was approached by a Spanish supporter of the insurgents who told him that Spain was in danger of being "reduced to a loathsome dependency of Soviet Russia with an anti-Christ Government. Surely the Christian countries of Europe could not possibly look with favour upon the prospects of Spain as a new outpost of Russia."[48] On 23 September 1936 O'Duffy traveled to Spain bearing the following message to the Nationalist leadership: "Ireland is behind the people of Spain in their fight for the Faith." He also wrote to the Dublin papers suggesting the formation of an Irish volunteer brigade. "I merely stated the issue at stake as it appeared to me," he wrote in his memoir, "that General Franco was holding the trenches, not only for Spain, but for Christianity."[49] As we shall see later on, the actual experiences of O'Duffy's Irish Legion in Spain fell short of the general's high expectations.

Obviously, the distinction between two diametrically opposed, homogenous, ideological blocks must not be oversimplified. There were divisions within the leftist camp in Spain, with some foreign volunteers serving with anarchist and Trotskyist units that ended up clashing with the Republican government. The British volunteer and later world-famous author George Orwell described his experience of in-fighting among pro-Republican forces in Barcelona as "one of the most unbearable periods of my whole life." He added that "few experiences could be more sickening, more disillusioning, or, finally, more nerve-racking than those evil days of street warfare."[50] Meanwhile, the Irish Catholics, White Russians, and members of the Romanian Iron Guard who fought for Franco, supposedly united in the defense of "Christian civilization," each had their own agenda for joining the conflict. Volunteers on the Right and on the Left were often mobilized more because of the political divisions in their own countries than because of a deep understanding of the issues at stake in Spain.[51]

When the Second World War broke out, the ideological aspects of the conflict attracted foreign volunteers from both neutral and occupied countries. The British armed forces incorporated tens of thousands of exiles from all over conquered Europe and also enlisted volunteers from neutral countries such as Ireland and, before December 1941, the United States.[52] Nazi Germany was initially very selective in its recruitment of foreigners. Because of its racial ideology, only so-called Germanic volunteers were allowed to join the Waffen SS. Originally the armed wing of the Nazi Party's protection squads, the Waffen SS morphed during the war into a military organization that competed with the German army, the Wehrmacht, over potential recruits. The head of the SS, Heinrich Himmler, began to recruit western and northern Europeans who were not citizens of prewar Germany in late 1940.[53] A recent detailed study into the prewar backgrounds of Danish, Swedish, and Swiss officers in the Waffen SS reveals that many of them had an ideological inclination toward fascism and aspired for a pan-Europeanist future. According to historian Martin Gutmann, these volunteers believed in the cultural and, in some cases, racial commonality of northern Europe, and they wished to see the continent reinvigorated:

> Spawned from their disgust with what they perceived as their countries' and the West's stagnation in culture, politics and economics, and the perceived threat from both the Bolshevik East and the Anglo-liberal West, they wished for a new form of organisation for what they saw as the heartland of European civilisation. Though not fully articulated in all these men, the seeds of such a world-view and of such a longing were present long before they donned the Waffen-SS uniform.[54]

The invasion of the Soviet Union in the summer of 1941 radically changed Germany's foreign recruitment practices. As war losses mounted and massive shortages in manpower became apparent, racial guidelines were relaxed. Both the Wehrmacht and the SS began to recruit troops from the supposedly inferior races of eastern and southeastern Europe. The Wehrmacht began recruiting among prisoners of war and from the Azerbaijani, Turkestani, Kalmyk, Ukrainian, Georgian, and Armenian civilian populations. Hundreds of thousands were recruited into the so-called eastern troops (*Osttruppen*) in this way. Most of the recruits were driven by material interests, such as better provisions in the case of prisoners of war, rather than belief in the ideas espoused by the Nazi regime. In terms of ideology, they had little in common with the Danish, Swiss, and Swedish volunteers who enlisted at the beginning of the war. The Waffen SS too began to recruit Crimean, Turkic, and Caucasian SS units in the East, and Bosnian and Albanian divisions in the Balkans. Eventually, approximately half a million of the soldiers of the Waffen SS were recruited outside Germany.[55]

On an ideological level, the Nazi leadership sought to rally western Europe behind them by depicting the conflict as a pan-European struggle against Bolshevism. Himmler hoped to lure a greater number of western European recruits into the Waffen SS by painting the struggle as a "Crusade against Bolshevism."[56] Nazi propaganda clung to this appeal until the bitter end. In early 1945, with the Red Army advancing toward Berlin from the east, German officials prepared a pamphlet in English for American, British, and Commonwealth soldiers:

> The fate of your country too is at stake. This means the fate of your wives, of your children, your home. It also means everything that make[s] life livable, lovable and honorable for you. . . . Extraordinary events demand extraordinary measures and decisions. . . . Soldiers! We are sure there are some amongst you who have recognized the danger of Bolshevik Communism for his own country. . . . We think that our fight has also become your fight. If there are some amongst you who are willing to take consequences and who are willing to join the ranks of the German soldiers who fight in this battle which will decide both the fate of Germany and the fate of your countries we should like to know it. We invite you to join our ranks and the tens of thousands of volunteers from the communist crushed and conquered nations of eastern Europe, which have had to choose between submission under an [sic] most brutal asiatic rule—or a national existence in the future under European ideas, many of which, of course, are your ideas. . . . Are you for the culture of [the] West or the barbaric asiatic East?[57]

The defeat of Nazism did not end the ideological struggle between Right and Left. The Cold War between the Eastern Bloc, led by the Soviet Union, and the West, led by the United States, created further ideological fault-line conflicts that attracted foreign volunteers. In fact, if one person symbolized the foreign volunteers of the second wave in the same way that Garibaldi epitomized the nineteenth-century struggle for liberty, it would be Ernesto "Che" Guevara. Growing up in Argentina, the young Ernesto followed the plight of the Spanish Republic on a map pinned to his bedroom wall. Furthermore, his parents founded a committee to help Republican refugees who arrived in Argentina. Like Garibaldi, Guevara would develop an expertise in guerrilla warfare and a taste for revolution rather than government. Like the "hero of two worlds," the young Argentine had an internationalist outlook and started his travels by setting out to sea. In 1951 he enrolled as a nurse with the Argentine merchant fleet and went on two trips that brought him to Brazil, Venezuela, and Trinidad and Tobago. In 1952 he embarked on his famous motorcycle journey across South America, during which his dream of creating a United States of Latin America, free from colonialism and offering equality of opportunity to all, began to take shape.[58]

Back in Argentina in 1953, he completed his medical studies but, restless as ever, he returned to his travels soon afterward. Guevara was in Guatemala in 1954 to support and learn from the social reforms of the leftist government of Jacobo Árbenz. When a force backed by the CIA staged a coup against the government, the young Argentine volunteered his services. "I myself have been assigned to emergency medical service and have also enrolled in the youth brigades to receive military instruction for any eventuality," he wrote to his mother.[59] He later described this period as "the ebb of democratic governments in 1954, when the last Latin American revolutionary democracy still standing in the area . . . succumbed to the cold, premeditated aggression carried out by the United States."[60] Once President Árbenz was toppled, Guevara moved to Mexico. It was there that he met Fidel Castro and became involved in the plan to stage a revolution in Cuba.

Being a foreigner among Cuban revolutionaries was not without its problems. Years later Castro recalled the time when he appointed Guevara as leader of one of the safe houses in Mexico City.

> There were about twenty or thirty Cubans there in all . . . and some of them . . . challenged Che's leadership because he was an Argentinian, because he was not a Cuban. We of course criticized this approach . . . this ingratitude towards someone who, although not born in our land, was ready to shed his blood for it. And I remember the incident hurt me a great deal. I think it hurt him as well.[61]

But after the revolutionaries reached Cuba in 1956 and began a protracted guerrilla campaign against the pro-US regime of Fulgencio Batista, Guevara was able to earn the respect of his peers. Intelligent, brave, at times even fool-hardy, he became one of the most prominent figures in the revolutionary movement. Appalled by the poverty and illiteracy of the local population, he oversaw the setting up of schools and clinics in the rebel-controlled Sierra Maestra. At the same time he also earned a reputation as a harsh discipli-narian, executing deserters and suspected spies.

After the revolution succeeded in toppling Batista, Guevara held a number of governmental positions and embarked on a few diplomatic voyages, mainly in Asia and Africa, but struggled to adapt to the life of a statesman. He con-tinued to champion a revolutionary philosophy that, by the 1960s, had two layers. First, he saw Cuba as the vanguard to the liberation of the whole of Latin America, which he saw as "a more or less homogenous whole, and in almost its entire territory U.S. monopoly capital holds absolute primacy." In the whole continent, he believed, "the puppets or . . . weak and timid gov-ernments are unable to resist the orders of the Yankee master." Second, and more broadly, he thought in terms of a conflict throughout the "fundamental field of imperialist exploitation [that] covers the three backward continents—Latin America, Asia and Africa."[62] In August 1960 he gave a speech in Ha-vana where he declared:

> Despite all that is said to the contrary . . . the forms of capitalism
> we have known, under which we have been raised and have suf-
> fered, are being defeated throughout the world. . . . And we have
> had the pride and self-sacrificing duty of being the vanguard in
> Latin America of a liberation movement that began some time ago,
> in the other subjugated continents of Africa and Asia.[63]

Guevara practiced what he preached. In April 1965 he wrote to Castro, re-nouncing his roles in the party leadership, his rank of comandante, and his Cuban citizenship.[64] Guevara returned to the life of a guerrilla leader, trav-eling with a group of Cubans to the Democratic Republic of the Congo with the aim of assisting the leftist Simba rebellion there. This mission was a com-plete failure. Guevara soon became frustrated with the Congo rebels' lack of discipline, corrupt and weak leaders, and failure to enlist the support of the local population.[65] While he "tried to show them that we are talking not of a struggle within fixed frontiers" but of a war against a common enemy across Africa, "no one saw it like that."[66] After the Cuban force was made to retreat and abandon Congo, Guevara found solace in the thought that "foreigners who went to risk their lives in an unknown land where people spoke a different language and were linked to them only by ties of proletarian internationalism, so that a method not practised in modern wars of liberation was thereby inaugurated."[67]

The failure in Africa did not dampen Guevara's desire to export the revolution to other countries. His next and final stop was Bolivia. Unlike Garibaldi, who endorsed and sometimes traveled to join existing conflicts abroad, Guevara wanted to spark an armed rebellion where none existed. Furthermore, he hoped that the revolution would spread from Bolivia to other South American countries. Before leaving Cuba, he wrote the "message for the Tricontinental," addressed to the Organization of Solidarity of Asian, African and Latin American Peoples. His message was a global one:

> Let the flag under which we fight be the sacred cause of the liberation of humanity, so that to die under the colours of Vietnam, Venezuela, Guatemala, Laos, Guinea, Colombia, Bolivia . . . will be equally glorious and desirable for a Latin American, an Asian, an African, and even a European. Every drop of blood spilled in a land under whose flag one was not born is experience gathered by the survivor to be applied later in the struggle for liberation of one's own country.[68]

Guevara believed that his guerrilla force should be made up of fighters from all over Latin America, but his main problem was enlisting the support of the Bolivian peasantry. Outnumbered, exhausted, and surrounded, his force was annihilated by Bolivian army units in October 1967. Guevara himself was injured, captured, and eventually executed.

The second wave of ideologically inspired foreign volunteers, fighting in fault-line conflicts between Left and Right, lost its momentum as a direct result of the end of the Cold War and the collapse of the Soviet Union. In the late 1980s and early 1990s, there were a few foreigners fighting alongside the Karen National Liberation Army against the socialist regime in Burma such as the Frenchman Jean-Philippe Courrèges. Some of these foreigners were from the Far Right and held staunchly anti-communist views.[69] During the 1990s there were still a few leftist volunteers—such as Andrea Wolf from Germany—who traveled to join the PKK in its armed struggle against the Turkish government.[70] By this stage, however, the ideological fault lines had shifted and a new wave of foreign volunteers began to emerge.

The Clash of Civilizations

In an influential though controversial *Foreign Affairs* article in 1993, Samuel Huntington famously argues that, in the post–Cold War era, the dominating source of conflict in the world would be cultural rather than economic or ideological. He believed that "the principal conflicts of global politics will occur between nations and groups of different civilizations" and predicted that the "fault lines between civilizations will be the battle lines of the future."[71] It is safe to assume that most of the foreign volunteers in the last twenty years have not read Huntington's work. Probably the majority of them never even heard of the Harvard professor.[72] However, in their global outlook, a great many foreign fighters, as they have come to be known in the post–Cold War period, have tended to think in Huntingtonian terms.

An important antecedent to the idea of a clash between rival civilizations was provided by Sheikh Abdullah Azzam. He was born in a village near Jenin in 1941, in what was then British-ruled Mandatory Palestine. He grew up in a devout Sunni Muslim family. Indeed, Azzam himself was deeply religious from a young age, joining the local branch of the Muslim Brotherhood, a political-religious organization that was founded in Egypt in the late 1920s. Azzam trained and worked as a teacher, teaching in schools in Jordan and the West Bank (which Jordan annexed after the 1948 war with Israel). In the Six-Day War of 1967, Israel conquered the West Bank. Deeply rattled and disappointed by the Arab defeat, Azzam moved his family to Jordan. In 1968 he left his teaching job to join the armed struggle against Israel. A number of Palestinian organizations used bases in Jordan to launch attacks against the Israelis. It was not long before Azzam was disturbed by the secular lifestyle of his peers and by the adherence of some of the Palestinian organizations to Marxism. After the Jordanian government clamped down on the Palestinian militias in what came to be known as "Black September" of 1970, Azzam set aside the armed struggle to pursue religious studies. He moved to Egypt where he gained a PhD from Al-Azhar University in Cairo in 1973. Returning to Jordan, he began to work as a university lecturer, a career he continued to pursue in Saudi Arabia and later in Pakistan in the early 1980s.[73]

One of the main reasons that brought Azzam to Pakistan was its proximity to Afghanistan. In late December 1979 the Soviet Union had dispatched the Red Army to support the tottering pro-Soviet Afghan government. This sparked a war that was so fierce that it displaced more than 6 million refugees. Azzam wanted to help the local insurgents, known as mujahideen. In late 1983 he resigned his university position so that he could dedicate himself entirely to assisting the Afghan struggle. In 1984 he established a "services bureau" in Peshawar to support in-coming foreign volunteers en route to join the war in Afghanistan. One of his main partners in this endeavor, and its chief financier, was Osama bin Laden, a 27-year-old Saudi son of a billionaire construction magnate. The bureau sought to prepare the volunteers, mentally and militarily, for the battlefield. Later on Azzam and bin Laden also set up training camps. Between 1986 and 1989 both men took an active part in fighting the Red Army and its Afghan proxies. Azzam also traveled extensively in the Middle East, Europe, and the United States to bring attention to the plight of the Afghan people, raise money for their cause, and recruit volunteers to fight alongside them.[74]

Azzam's propaganda, and particularly his theological reinterpretation of the concept of jihad, would have a long-term impact on the ideology of Muslim foreign volunteers in the third wave. There are different kinds of jihad (struggle) in Islamic tradition, and each one of these has had a number of interpretations over the centuries. Azzam focused on jihad against disbelievers in the defense of Muslim lands or territories formerly held by Muslims. Starting with a fatwa (religious ruling) he published in 1984 and through his sermons, speeches, and the magazine he founded, *Al-Jihad*, he consistently emphasized the urgency of joining the Afghan jihad.[75] "Anybody who looks into the state of the Muslims today will find that their greatest misfortune is their abandonment of Jihad," he wrote in his recruitment booklet, *Join the Caravan*.[76] Unlike other religious Islamic scholars, who believed that fighting non-Muslim invaders was an obligation only for the

Muslims living in the country under attack, Azzam saw it as the personal obligation of all Muslims, no matter where they lived.

> When the enemy enters the land of the Muslims, Jihad becomes individually obligatory (Fard Ain). . . . When Jihad becomes Fard Ain, no permission of parents is required. . . . Donating money does not exempt a person from bodily Jihad, no matter how great the amount of money given. Nor is the obligation of Jihad, which is hanging over the person's neck, lifted from him. . . . Jihad is currently Fard Ain—individually obligatory—in person and by wealth, in every place that the disbelievers have occupied. It remains Fard Ain continuously until every piece of land which was once Islamic is regained.[77]

For Azzam, the Afghan jihad was a matter of great urgency. As he explained in a lead article in *Al-Jihad* in March 1985, "We missed the opportunity in Palestine and now it may be lost in Afghanistan. The historical responsibility is great and time is not on our side."[78] Azzam therefore saw all the Muslims of the world as one community, the *umma,* paying no heed to national boundaries. He also assumed that all the Muslims whose country had come under non-Muslim rule wanted to be liberated (much like Che Guevara assumed that all the peoples of Africa and Latin America wanted to see their capitalist governments overthrown). Azzam's doctrine created quite a commotion in Islamic circles. His interpretation of jihad was accepted by some notable Islamic scholars, especially in Saudi Arabia, but was rejected by many others.[79] Regardless of whether they were directly influenced by Azzam's preaching or not, the number of foreign volunteers who traveled to join the fight against the Soviets in Afghanistan grew in the second half of the 1980s. Although numbers are heavily contested, with estimates varying between 5,000 and 25,000, the "Afghan Arabs," as they were sometimes called, became the largest foreign volunteer movement in decades.[80] Azzam, in many respects the father of this movement, was assassinated in Peshawar in November 1989. The culprits were never identified.

Abdullah Azzam and the war against the Soviet Union in Afghanistan created the ideological infrastructure for volunteering to wage jihad abroad against non-Muslims, but at this stage the "clash of civilizations" had not yet begun in earnest. After all, the US government supported the Afghan cause, providing the mujahideen with both money and weapons. Western popular culture too saw this conflict through the lens of the Cold War, with both James Bond and Rambo lending a symbolic hand in the Afghan fight against the Soviet Union.[81]

However, the end of the Cold War complicated the geostrategic horizon. The conflict in the former Yugoslavia in the early 1990s left Western governments baffled. Media coverage of developments in the Balkans alongside what many perceived as international inaction attracted several hundred volunteers to fight in this conflict, on different sides. Once a federal socialist state, Yugoslavia began to break up in 1991, with independence declarations from two of its constituent republics: Slovenia and Croatia. Croatian independence was resisted with force by some of the country's Serbian minority and also by the Serbian-dominated Yugoslav National Army. In 1992 Bosnia and Herzegovina, another constituent republic, declared its independence; a declaration that immediately pitted the predominantly Bosnian-Muslim government against the Bosnian-Serb minority and their patrons in Belgrade. Soon Bosnian Croats also joined the fray. Bosnia descended into a brutal war that lasted until 1995.

Foreign volunteers fought alongside nearly every armed faction in the Yugoslav Wars, a conflict that Huntington described as a "fault line war" that pitted what he called the Western, Orthodox, and Islamic civilizations against each other.[82] The armed forces of Republika Srpska, the Bosnian-Serb entity, received help from approximately 500 to 700 Russians, some 100 Greeks, and a few other foreign nationals.[83] As in so many other cases, these fighters emerged against a backdrop of broader support for the Serbs in their countries of origin. The Greek media was largely supportive of the Serb cause. Meanwhile, the Russian writer and leader of the National Bolshevik Party, Eduard Limonov, was filmed on the hills above Sarajevo, accompanied by Bosnian-Serb leader Radovan Karadzic, firing a machine gun into the besieged city.[84] As we shall see later on, a few hundred volunteers, mainly from western, central, and northern Europe, joined Croatian units and fought both in Croatia and in Bosnia.[85] And then there were volunteers from predominantly Muslim countries, approximately 1,000 according to a recent estimate, who fought alongside the Bosnian Muslims.[86]

Scholars have highlighted, and at times even overemphasized, the role played by foreign veterans of the Afghan conflict in the war in Bosnia.[87] Many of the Afghan Arabs returned to their home countries after the Soviet withdrawal from Afghanistan and the death of Abdullah Azzam in 1989. Others stayed on to fight against the pro-Soviet government, still in power in Kabul, or to join Al-Qaeda, the secret movement formed by Osama bin Laden. In variance with Azzam's doctrine of mobilizing in defense of Muslims wherever these were under attack by non-Muslims, bin Laden and the predominantly Egyptian militants around him went one step further. They wanted to overthrow those Muslim governments whom they perceived as "apostate" and, ultimately, to wage war against the so-called Crusader-Zionist alliance (the West, led by the United States, and Israel).[88]

From this point onward, some militant Muslims would follow the path set by Azzam and volunteer to fight in defense of fellow Muslims; some would follow the path of bin Laden, which increasingly involved international terrorist activities; and some would do both. Indeed, at times the two forms of militancy fed each other. The involvement of veterans of the Afghan conflict in terrorist attacks in Egypt, Algeria, and on the World Trade Center in New York in 1993 led to pressure on the Pakistani government to halt the activities of the Arab militants that still resided in the country. Hence, the Pakistani authorities exerted pressure on the Arabs in Peshawar to leave and, subsequently, a few dozen veterans of the struggle in Afghanistan went on to fight in Bosnia.[89] These veterans were joined by a new generation of foreign volunteers, mainly from North Africa and the Middle East.

For some of the volunteers, participation in this war was about helping fellow Muslims. A recruitment video in which two British Muslim volunteers in Bosnia were interviewed echoed Azzam's doctrine. A 21-year-old medical student from Birmingham University, who identified himself as "Abu Ibrahim," declared:

> In the West many brothers say to us that the Muslim *umma* needs doctors, they need lawyers, they need scientists, they need engineers. And I disagree with that because there is enough Muslim doctors, there's enough lawyers, scientists, engineers. But what we lack here is Muslims that are prepared to suffer and sacrifice.[90]

The Yugoslav Wars also coincided with an anti-Western shift in the jihadists' discourse. Looking back at his time in Bosnia, Saudi-born volunteer Aimen Dean recalled how, at the end of the conflict, "those [comrades] who survived started to adopt a far more anti-Western feeling, that the global community were conspiring against the Muslim community." This was because, in late 1995, the tide of the war was turning in the Bosnian Muslims' favor, and so the West "wanted to end the war there and then before they [Bosnian Muslims] scored any more victories." To this he added: "I think they started to feel that the West is fighting Islam as a religion and that there is an anti-Muslim conspiracy being formed in Washington and London and Paris and that led to further radicalization."[91]

The leadership of Al-Qaeda saw the plight of Muslims in Bosnia and in other conflict zones—such as Chechnya—as part of a broader crisis of the Muslim world. In 1996 Osama bin Laden returned to Afghanistan and soon declared that he was at war with the United States. He argued that, because of the United States and its allies,

> the Muslims' blood became the cheapest and their wealth as loot in the hands of the enemies. Their blood was spilled in Palestine and Iraq. . . . Massacres in Tajikistan, Burma, Kashmir, Philippines, Somalia, Eritrea and Chechnya and in Bosnia-Herzegovina took place, massacres that send shivers in the body and shake the conscience.[92]

Bin Laden had good reasons to highlight Muslim suffering. As we have seen, vicarious plight makes for very potent recruitment propaganda. Recent studies have shown that militant Muslims were far more likely to mobilize to fight in defense of co-religionists abroad than they were to join or even support terrorist activities in their home countries.[93] As one radical British Muslim explained:

> I understand the acts of resistance in Afghanistan. They are freedom fighters. They didn't go to the United States to carry out acts of terrorism. I sympathise with the people who fight in Afghanistan, in Iraq, in Palestine, in Sudan. I don't agree with the ones who commit acts in an underground, a train, a plane.[94]

Al-Qaeda and its affiliates were nonetheless able to benefit from the willingness of militant Muslims to fight for their co-religionists abroad. A small percentage of foreign fighters have returned to their countries of origin to take part in terrorist plots, as we shall see in Chapter 7. Others initially mobilized to become foreign fighters, attended Al-Qaeda's training camps in Afghanistan, and ended up getting involved in plans to carry out attacks in the West instead.[95]

The third wave of fault-line conflicts attracting foreign volunteers appears to be in full strength in the early twenty-first century. The protracted conflict in Iraq, which began after the American-led invasion of 2003, summoned between 4,000 and 5,000 foreign volunteers. The struggles against foreign invaders waged by the Taliban in Afghanistan since 2001 and Al-Shabaab in Somalia since 2006 also drew fighters from abroad, though in smaller numbers.[96]

The bloodletting in Iraq paved the way for a major development in the ideological discourse surrounding foreign volunteers in conflicts that involved Muslims. Here, a central role was played by the Jordanian Abu Musab al-Zarqawi. Hutaifa Azzam, the son of Abdullah Azzam, described Zarqawi as a violent "street guy" who had been drinking alcohol before "he finally decided to return back to Allah."[97] In the early 1990s Zarqawi traveled to Afghanistan to fight and spent time in a Jordanian prison in the years that followed. He was back in Afghanistan in 2001, when the war against the United States–led coalition began, later making his way into Iraq. In late 2004 he pledged the allegiance of the organization he had formed to Al-Qaeda.

In some respects Zarqawi's Al-Qaeda in Iraq echoed the doctrine of Abdullah Azzam, seeing jihad against the non-Muslim invaders as an obligation for all Muslims. Zarqawi's desire to use Iraq, once liberated, as the basis for a revived Islamic caliphate also has its roots in Azzam's doctrine. In the 1980s Azzam had believed that the "establishment of the Muslim community on an area of land is a necessity, as vital as water and air." The establishment of such a "homeland" was one of the long-term goals of the jihad he advocated.[98] Zarqawi's innovation was that he sought to achieve his goal not only through fighting the Americans but also by launching attacks against the Shia population in Iraq. In the 1980s Sunni and Shia Muslims found a common enemy in the Soviets who had invaded Afghanistan. In the following decade bin Laden and Al-Qaeda steered clear of stoking the flames of historic Sunni–Shia enmity. In contrast, Zarqawi believed that attacks against the Shia majority in Iraq would spark a civil war that would galvanize the country's (and the region's) Sunni population, spurring them into joining the jihad.[99] In this way the modern clash of civilizations kindled a much older sectarian clash, dating back to the seventh century.

The civil war in Syria that began in 2011 has already eclipsed previous conflicts in the third wave in terms of the number of foreign volunteers it has attracted. From the point of view of Abdullah Azzam's doctrine, this seems odd, as initially there was no foreign invader in this war. Motivations for volunteering to fight in Syria are varied and complex, much like in most of the other conflicts discussed thus far. The plight of the Syrian population certainly produced a lot of sympathy, especially in sectors that were inimical toward Bashar al-Assad's secular Baathist regime. In terms of geopolitics, the alliance between Syria's Alawite ruling elite, the Lebanese Shia organization Hizballah, and the Shia regime in Iran has given an incentive to those who seek to defend fellow Sunnis in Syria.

One Sunni organization that has flourished in the conditions created by the civil war in Syria is the self-proclaimed Islamic State, initially an off-shoot of Zarqawi's Al-Qaeda in Iraq. The notion of a global jihad against disbelievers is central to its ideology. In a speech he delivered in May 2015, the organization's leader, Abu Bakr al-Baghdadi, synthesized Abdullah Azzam's message, that Muslims have an obligation toward their co-religionists, and bin Laden's belief in an uncompromising war against Christians and Jews:

> O Muslims! Do not think that the war we are waging is the Islamic State's war alone. Rather it is the Muslims' war altogether. It is the war of every Muslim in every place, and the Islamic State is merely the spearhead in this war. It is but the war of the people of faith against the people of disbelief. . . . March forth everywhere, for it is an obligation upon every Muslim who is accountable before Allah. And whoever stays behind or flees, Allah (the Mighty and Majestic) will be angry with him and will punish him with a painful torment.[100]

If the Islamic State goes further than its predecessors in terms of ideology, it is in the emphasis it places on a dual obligation of jihad and *hijrah* (emigration). After all, the organization seeks not only to defeat its enemies but also to create a new political entity. In the words of al-Baghdadi, "there is no excuse for any Muslim who is capable of performing hijrah to the Islamic State, or capable of carrying a weapon where he is, for Allah . . . has commanded him with hijrah and jihad, and has made fighting obligatory upon him."[101] For this reason the Islamic State goes beyond recruiting male foreign fighters and encourages women and sometimes even families with young children to come to territories under its control in Syria and Iraq.

The brutality of the Islamic State has shocked many, not only in the Middle East but in more distant countries too. This brutality prompted volunteers from various countries in the West to fight alongside the YPG, a Kurdish force that has been playing a key role in the war against the Islamic State in Syria. According to some estimates, there were approximately 100 foreign volunteers fighting alongside the Kurds in summer 2015.[102] One of them was John Gallagher, a former Canadian light infantry soldier from Ontario who was killed in November of that year.[103] A few months earlier, Gallagher explained his reasons for traveling to the Middle East to fight on his Facebook page: "This war is about ideas as much as it is about armies. Slavery, fascism, and communism were all bad ideas which required costly sacrifice before they were finally destroyed. In our time, we have a new bad idea: Theocracy."[104] He believed that the war against theocracy started with the Iranian Revolution of 1979, if not earlier, and that the "Muslim world has been dominated by theocratic politics for decades now." He also argued that "only by destroying ISIS without mercy can we discredit the idea, and force the would-be jihadis and fellow-travelers to give up their insane dreams." He concluded by saying: "I'm prepared to give my life in the cause of averting the disaster we are stumbling towards as a civilization." Judging from the duration of the first and second waves that attracted foreign volunteers, the clash of civilizations wave may well continue for a number of decades before losing its momentum.

* 3 *

A Search for Meaning

Deciphering Motivations

The Motivation Enigma

Abd al-Aziz al-Muqrin, from Riyadh in Saudi Arabia, traveled to Afghanistan in 1990 and fought against the forces that remain loyal to President Mohammad Najibullah after the withdrawal of the Soviet Union. A few years later he took part in the war in Bosnia. By the early 2000s he had become a prominent figure in Al-Qaeda.[1] In an interview he gave in 2003, he claimed that few Saudi boys coming of age in the 1980s were able to resist the desire to fight against invading Soviet forces in Afghanistan. Indeed, the Afghani struggle was praised in the mosques and by the local media.[2] However, according to the most generous estimates, the number of Saudis who joined the jihad in Afghanistan during that decade was 4,000 to 5,000.[3] In the 1980s the population of Saudi Arabia passed the 10 million threshold. Hence, only a tiny fraction of young Saudis traveled to Afghanistan while the overwhelming majority remained at home, despite the popularity of the Afghan cause and feelings of solidarity toward oppressed fellow Muslims.

This was not the only case where the pool of potential supporters was far larger than the actual number of foreign volunteers. The conflict between Jews and Arabs in Palestine in late 1947 and early 1948 had an immense impact on Jewish communities in the United States. In a number of overt and secret fund-raising campaigns, vast amounts of money were collected for the Zionist cause. The United Jewish Appeal, for instance, raised $150 million in 1948, four times more than the American Red Cross's entire annual campaign. A large number of American Jews also helped locate surplus army machinery for manufacturing guns and bullets while Zionist youth volunteered to package munitions in deceptively marked crates. Their willingness to support what soon became the State of Israel cannot be questioned. Yet, from a population of approximately 5 million Jewish Americans, only a few hundred volunteers traveled to the Middle East to take part in the war.[4] What distinguished those who went from those who stayed behind? Clearly, solidarity, ideology, and religious beliefs on their own do not provide a sufficient explanation for the phenomenon of foreign volunteers.

As historian Michael W. Jackson has pointed out, "volunteering is a question of motivation, which falls within the realm of the emotions, of the subjective."[5] Assessing motivations is a challenge, and doing so based on personal accounts is problematic. Retrospective accounts in particular are subject to omissions, self-censorship, and other shortcomings, as the following examples illustrate. One such problem is over-embellishment and unreliability. There is no doubt that John MacPhee from Scotland was in Bosnia-Herzegovina in the 1990s, and that he had joined a Bosnian-Croat unit during the Yugoslav Wars. However, while his memoir, *The Silent Cry*, could undoubtedly furnish scenes for an action movie filled with daring commando raids behind enemy lines, fellow volunteer Rob Krott describes the book as "full of fanciful stories and lies."[6]

The autobiography *Dual Allegiance*, written by the Jewish-Canadian volunteer Ben Dunkelman, who served in the Israel Defense Forces (IDF) in the war of 1948, provides a good example of self-censorship. In July 1948 Dunkelman was in command of 7th Brigade, which captured the city of Nazareth. When Dunkelman was ordered by his superior to expel the city's Arab population, he refused. Though he was soon removed from the position of military governor of the city, Dunkelman's stance against expulsion was approved by the prime minister and minister of defense, David Ben-Gurion (the Israeli leader was less accommodating when other Arab towns were captured by the IDF). However, the page describing this chain of events was omitted from the final version of Dunkelman's autobiography, which was published in the late 1970s. As a pro-Israeli veteran, he must have wanted to avoid embarrassing the Israelis, whose narrative claimed that there was no policy of expelling the Palestinian Arabs in 1948. The omitted page came to light in a left-wing magazine that sought to challenge the official Israeli position and was probably made available by Dunkelman's ghostwriter.[7] Indeed, volunteers in a number of conflicts have felt the need to adjust their retrospective accounts so that these would fit into broader political narratives. A few veterans of the Spanish Civil War's International Brigades, for instance, have published memoirs that gloss over events that would have painted the communist movement in a negative light.[8]

Another well-known problem with retrospective personal accounts is the ability of authors to accurately remember what happened. Flora Sandes, a British volunteer in the Serbian army during the First World War, candidly admitted, "My memories of the 1916 campaign are confused. They seem like a whole series of vivid pictures of little incidents which I can never forget, but which are not consecutive."[9] And then there is the almost unavoidable tendency that people have to reinterpret the actions and decisions of their younger selves through the lens of the knowledge and experiences they have accumulated since. When David Karon was asked in an oral history interview in Israel in 1965 about the reasons that led him to volunteer to fight in the Spanish Civil War he replied:

> Well, this is actually a question which is a bit hard to answer. It would have been much easier to answer had you asked me then. Because today, after nearly thirty years have passed, it is difficult to say what the reasons were without, perhaps, adding an element of life experience gathered over thirty years.[10]

In addition to these methodological hurdles that historians face, there is the broader problem of determining the exact reasons that prompt individuals to make decisions. In the words of Graciela Iglesias Rogers, "attempting to establish the precise motivations behind any human action is inevitably difficult."[11] As studies in psychology have been arguing for quite some time, human beings are far less rational in their decision making than we might think.[12] However, despite these difficulties, the attempt to understand the motivations of foreign volunteers should not be seen as a lost cause. The task of the historian is to identify patterns and repeat occurrences as a means of overcoming the shortcomings that afflict individual sources.

One discernible, recurring pattern is the way in which the decision to travel abroad to fight could be motivated by a mixture of political, social, moral, and personal reasons. These motivations were, and are still, frequently interwoven in such a way that is difficult to untangle. To give but one example, Peter Kemp—one of the few British volunteers who joined the Nationalists and fought alongside Franco's forces in the Spanish Civil War—rationalized his motivations for going to Spain by giving an array of reasons. On the one hand, there was "a feeling for adventure, the desire to be on my own and independent." He also wished "to see a new country, get to know the people and learn to speak their language." On the other hand, Kemp "felt strongly, too, that as a Tory who had spoken frequently and at length in debates at the Cambridge Union I should be prepared now to stick my neck out in support of my views." Indeed, he harbored "a deep-seated hatred and fear of communism and what it might do to Britain and Europe." In his memoir he confessed that "even today I'm not sure which of them [motivations] was the strongest."[13]

Foreign volunteers are attracted to participate in certain conflicts, but their decision making is also influenced by their inner and surrounding predicaments. It is a key contention of this book that, if foreign volunteers from different backgrounds, different ideological inclinations, and, indeed, different historical periods have one thing in common, it is their search for a sense of purpose and meaning. Fighting in a distant land could be understood as an attempt either to escape from a bewildering present or to fill some sort of void in the prewar lives of the volunteers. By choosing to fight in a conflict abroad, the volunteers seek to improve their self-esteem and gain recognition from their peers.

Agency

The *Daily Mail,* a right-wing British newspaper that was hostile to the Spanish Republic in the late 1930s, described volunteers in the International Brigades either as mercenaries or as dupes. It argued that these men were lured by promises of safe work behind the lines, only to be dragooned into active service once they arrived in Spain.[14] Meanwhile, supporters of the Republican cause depicted the foreign volunteers as a spontaneous, grassroots movement. Herbert Greene (brother of the author Graham Greene), who visited

Spain during the war, argued that "they are all keen volunteers and the stories in the English papers that they are inveigled into going by false pretences are untrue."[15] This wartime wrangling continued into the Cold War. In 1958 Federal Bureau of Investigations Chief J. Edgar Hoover described members of the International Brigades as "dupes." For him, a dupe was an "individual who unknowingly is under communist thought control and does the work of the Party."[16] Traces of this dispute can be found in the partisan historiography of the Spanish Civil War.[17] In 1982 American historian R. Dan Richardson depicted the foreign volunteers as pawns in an international communist conspiracy: "without the Soviet-Comintern decision to intervene directly in the Spanish War . . . the International Brigades would never have come into existence."[18] Conversely, Michael W. Jackson argues that analyzing the Comintern's activities during the war "only tells us how and not why they went. At most, the organizational efforts of communism harnessed forces it did not generate."[19]

Debates about the extent to which foreign volunteers have had much agency in deciding to travel abroad to fight are not restricted to studies of the Spanish Civil War. Similar arguments have been raised in contemporary discussions about the radicalization of Muslim foreign fighters.[20] The possibility of manipulation becomes more acute the younger the recruits, as in some cases where Islamic State members have "groomed" male and female teenagers online.[21] In a recent attempt to identify structural similarities between different historical instances in which foreigners have engaged in wars abroad, David Malet claims that "insurgencies try to recruit foreign fighters by framing distant civil conflicts as posing a dire threat to all members of a transnational community of which both the foreign recruits and local insurgents are members."[22] According to this approach, "recruiters manipulate identities to make them salient" and "prime their audiences before pitching participation by describing the significance of the conflict." For their part, the fairly passive foreign volunteers internalize "the narratives of the conflict promoted by recruiters."[23]

The evidence suggests that, while would-be foreign volunteers certainly could be influenced, the process was—and still is—more complex than the indoctrination model posits. To begin with, the decision to volunteer could often appear to be quite spontaneous. In a number of cases, an image in a newspaper or on a poster, a report on the television, and even a sentence in a conversation were sufficient to trigger the desire to enlist to fight. For instance, Fred Thomas from London had the following to say about why he went to fight in Spain in 1937:

> I think emotion rather than cold reasoning determined the moment of decision. In the *Daily Express* I saw a picture of Franco's troops entering a captured village. Distraught women ran forward, arms outreached in helpless submission, pleading for the lives of their men. I could not bear the anguish in their eyes.[24]

Howard (Chaim) Cossman, a Canadian Jew who had served his country during the Second World War, described his decision to go to fight in Palestine in early 1948 thus:

> Having seen the picture of a Jewish girl with a rifle on her shoulder on the back page of the "Montreal Herald," I decided that if this is all they've got, I'm going to help. So I left my job as a salesman and returned to military life.[25]

Twenty-one-year-old "Abu Ibrahim," the British-Muslim volunteer in Bosnia who was interviewed for a recruitment video, used the following words to express his motivations:

> I watch the TV [in Britain] and tears role down my face when I see the Muslims in Bosnia, Muslims in Palestine, Muslims in Kashmir, and then I come here and you feel a sense of satisfaction, you feel that you are fulfilling your duty.[26]

The images of torture and prisoner abuse that emerged from the coalition-run prison at Abu Ghraib in Iraq in 2004 were cited by several Saudi recruits as catalysts for their decision to depart for the Iraqi battlefield.[27] In 2015 a 40-year-old British volunteer named Jim, fighting alongside Kurdish forces against the Islamic State in Syria, told Channel 4: "What really jolted me was the photograph I saw on Facebook of an ISIS fighter holding up the severed head of a woman. It seemed like one of the most evil, single images that I have ever seen . . . that affected me quite a lot."[28]

Few would doubt that the media helps to shape its audience's worldview. Conversely, very few would argue that the media at large sets out to produce more foreign volunteers. This was certainly not the objective of either the

Daily Express or the *Montreal Herald* in the examples mentioned above. Even in rare cases where encouraging enlistment was the specific aim of a media outlet—such as the magazine *Al-Jihad,* edited by Abdullah Azzam in the 1980s, or contemporary social media forums that seek to recruit volunteers for the conflict in Syria—the question still remains: why have some consumers decided to take up arms while others did not? To answer this question, we must examine the prewar outlook of the volunteers as well as their personal circumstances to see what role these played in how individuals processed information relating to the conflict abroad and how they decided to respond to it.

Foreign volunteers have a recurring tendency to interpret conflicts abroad through the prism of the political and social confrontations of their home societies. This reinterpretation could be done by groups that share backgrounds, layers of identity, and beliefs, or it could be done by individuals. The response of leftist black Americans to the Spanish Civil War provides an interesting case in point.

In late 1935, several months before the war in Spain erupted, many black Americans were enraged by Italy's invasion of Ethiopia (or Abyssinia, as it was often referred to at the time). Ethiopia was a symbol of African resistance against imperialism, having successfully repelled an Italian attack in 1896. It also commanded respect in some circles for being one of the first countries to embrace Christianity. Although some leftist black Americans criticized Ethiopia for not abolishing slavery, when the war with Italy broke out, there were several manifestations of black solidarity. A number of organizations were set up in large cities in the United States to raise money and support for the Ethiopian cause. In Boston, Chicago, Detroit, Los Angeles, and New York, some members of the black community even rioted and looted shops owned by Italian Americans. Organizations such as the Pan-African Reconstruction Association went one step further and attempted to recruit volunteers for Emperor Haile Selassie's army. Thousands of black Americans expressed an interest. However, these attempts came to nil because of the objection of the US government, the difficulty of ferrying volunteers to Ethiopia, and the relative brevity of the conflict (Mussolini declared victory in early May 1936, after his forces entered the Ethiopian capital, Addis Ababa).[29]

Some of the fervor that went into support for Ethiopia was transferred to the Republican cause in Spain after the outbreak of the civil war there.

The Communist Party in the United States depicted the war in Spain as an extension of the Italo-Ethiopian conflict. By early 1937 it adopted the slogan "Ethiopia's fate is at stake on the battlefields of Spain" and asked for material aid that had been collected for Ethiopia to be transferred to Spain instead. Approximately ninety black American volunteers joined the International Brigades. The poet Langston Hughes, who visited the volunteers in Spain, observed that by fighting against Franco, "they felt that they were opposing Mussolini."[30] One of these volunteers was 31-year-old James Yates. Born in Mississippi, Yates moved to Chicago and then to New York. In his memoir he described his deliberations about whether he should join his flatmate and volunteer:

> I had been more than ready to go to Ethiopia, but that was different. Ethiopia, a Black nation, was part of me. I was just beginning to learn about the reality of Spain and Europe, but I knew what was at stake. There the poor, the peasants, the workers and the unions, the socialists and the communists, together had won an election against the big landowners, the monarchy and the right-wingers in the military. It was the kind of victory that would have brought Black people to the top levels of government if such an election had been won in the USA.[31]

Yates saw a clear connection between Hitler, Mussolini, Franco, and racist politicians in the south of the United States. For him, the conflict in Spain was part of a broader battle against both racism and oppression. Similar views were expressed by Salaria Kea, who was a volunteer nurse in Republican Spain. A pamphlet distributed in her name by the Negro Committee to Aid Spain in 1938 argued that "Ethiopia's only hope for recovery lies in Italy's defeat. The place to defeat Italy now is in Spain." Like Yates' memoir, the pamphlet ties together "the lynching of Negroes in America, discrimination in education and on jobs" with the fate of the poor peasants of Spain, who "were so accustomed to poverty and hardship." The cause of Spain was also "the cause of minority groups throughout the world."[32]

Leftist black Americans were not the only group to interpret international events through the lens of their own social, cultural, and political realities. Communist miners from southern Wales saw a clear link between their struggle to protect trade unionism and the Spanish Civil War. Speaking at the Trades Union Congress of 1938, Will Paynter of the South Wales Miners' Federation told his audience: "I feel that the outcome of the war in Spain will not only determine the circumstances of our political existence, but will ultimately determine . . . the industrial existence of our trade union movement."[33] Many Jewish volunteers in the International Brigades saw the war in Spain in terms of a fight not only against fascism but also against antisemitism more broadly. Hyman Katz from New York, for instance, drew a link between the Spanish Inquisition of previous centuries, Hitler's persecution of the Jews, and Franco. Writing from Spain to explain his motivations for enlisting to his mother, he asked, "don't you realize that we Jews will be the first to suffer if fascism comes?" His response to this rhetorical question was unambiguous: "So I took up arms against the persecutors of my people—the Jews—and my class—the Oppressed."[34] Much like James Yates and Salaria Kea, Katz and many others like him were able to take a foreign cause and make it their own.

Although in the case of black American volunteers in the International Brigades group dynamic appears to have been dominant, in many other instances individuals have reinterpreted information regarding conflicts abroad independently, without being influenced by recruiters. One such example is the decision of Steve Gaunt, from Selby in Britain, to volunteer his services to fight for Croatian independence in November 1991. Gaunt, 33 years old at the time, was not a member of any political party. "I had my own ideas about things," he would later say, defining himself only as an "anti-communist."[35] He worked as an overseas operations manager for a holiday company. With the Yorkshire autumn in full swing, he found that there was not much to do in the office back in Britain. "My restlessness is increasing now the season is well and truly over," he wrote in his diary on 1 November 1991.[36] At the same time he was following the news about the growing conflict in Yugoslavia with great interest and was concerned about the fate of Croatia. Then, on 4 November, he "woke this morning and somehow decided to go to Croatia. Not knowing what to expect, I withdrew some cash and bought some good boots."[37]

The circumstances of another British volunteer for the Croatian cause, Simon Hutt, were different, though his decision too was not the result of pressure applied by recruiters. At 19, he was a soldier in the Royal Artillery and served with his regiment in the Gulf War of 1991. Somewhat confused by his experiences in the Persian Gulf and then the sudden return to the monotony of regimental life, he began to drink and use drugs. After he was caught and sentenced to thirty days' detention, he realized "that any career in the British Army that I may have had was now gone." Like Gaunt, Hutt was following events in Yugoslavia and was moved by the plight of the besieged Croatian town of Vukovar. He decided to go absent without leave and offer his services to the Croatians: "I could make a difference and for once in my life I would be standing up and doing the right thing." He was also enthusiastic about going "back into the intensity of a war."[38]

Like their predecessors, present-day would-be Muslim foreign fighters adjust the information they receive about conflicts in the Middle East so that it fits their own worldview. In the aftermath of the 9/11 attacks in New York and Washington, DC, some analysts have tended to focus exclusively on organization leaders and the ideals they promote in their propaganda. For instance, Mary Habeck argues that religion is the key to understanding the actions of jihadists: "Western scholars have generally failed to take religion seriously. Secularists, whether liberals or socialists, grant true explanatory power to political, social, or economic factors but discount the plain sense of religious statements made by the jihadists themselves."[39] When applied to foreign war volunteers, this approach privileges statements made by Abdullah Azzam, Osama bin Laden, and Abu Bakr al-Baghdadi about jihad and the religious duty to fight against non-Muslims. However, there are limits to the explanatory power of the "clash of civilizations" rhetoric, discussed in the previous chapter. In fact, some foreign fighters may have a very limited understanding of Islam. Take, for example, the court case of Yusuf Sarwar and Mohammad Ahmed, both 22, from Birmingham in the United Kingdom. They were arrested when they returned to Britain in January 2014, after spending several months in Syria, and charged with planning terrorist acts. During the proceedings it transpired that, as part of their preparations before leaving for Syria, they had ordered books online such as *Islam for Dummies* and *The Koran for Dummies*.[40] Lydia Wilson, a researcher who has interviewed imprisoned Islamic State fighters, is correct in observing that while Western volunteers joining the Islamic State can be deeply committed from a religious point of view, "it's to their own idea of jihad rather than one based on sound theological arguments or even evidence from the Qur'an."[41]

Foreign volunteers often see the conflict they wish to join in simplified terms. There is nothing exceptional about this tendency, for as psychologist Daniel Kahneman points out, "you will often find that knowing little makes it easier to fit everything you know into a coherent pattern."[42] But because of such a simplified approach, foreign volunteers may develop ideas that are not entirely compatible with those of their hosts. For instance, in March 2015 an unnamed British former soldier was interviewed by BBC Radio 4 about his intention to join Kurdish forces fighting against the Islamic State. When describing his reasons he said: "Look at Salah al-Din . . . many many years ago, he roamed all over Asia—didn't he—to spread his violence and terror, terrorizing people. Don't you think that's what ISIS is doing?"[43] One should hope that, if this volunteer managed to join Kurdish forces, he did not repeat this reason in conversations with his hosts. After all, Salah al-Din was of Kurdish extraction.

To summarize, would-be volunteers do not make the decision to volunteer in a vacuum. They were (and are still) susceptible to influence from social superiors (whether in a political party or as part of a religious hierarchy), peers, and the media. The clusters of International Brigaders from southern Wales and the Glasgow area in Scotland provide evidence not only for a collective interpretation of world events but also of peer pressure, with group members egging each other along.[44] Furthermore, foreign volunteers very often had to rely on the help of others to reach the warzone. As we see in Chapter 5, transnational networks of support played an extremely important role in ferrying foreign volunteers to the seats of war, especially in cases where large numbers were involved. But, even if the volunteers' view of the conflict was or is blinkered, they still exercise a degree of choice. Thus, the focus on "recruiters" only tells part of the story. It helps to explain *how* volunteers come to fight in foreign wars, but not *why*. Furthermore, as the following section shows, their personal circumstances when they decide to embark on foreign military service and the timing in which the opportunity to do so presents itself is of the utmost importance.

Push and Pull

In 1809 Spanish patriots were at war with Napoleon's forces. A few dozen volunteers from Britain offered their assistance to provisional Spanish authorities and ended up serving in various capacities. At the same time, the 21-year-old Lord Byron passed through Spain as part of his Grand Tour. "I like the Spaniards much," he wrote from Gibraltar in August 1809, "I should have joined the army, but we have no time to lose before we get up the Mediterranean & [the Greek] Archipelago."[45] His adventurous spirit was clearly already evident in 1809, but the circumstances were not yet ripe for Byron to commit himself to becoming a foreign volunteer. By the summer of 1823, the same Byron, who had been living in self-imposed exile for six years, was ready to take up arms for a foreign cause. Having recently moved to Genoa, leaving behind an extravagant lifestyle, his spirits were parched. One acquaintance reported that he "frequently said that the public must be tired of his compositions, and that he was certainly more so."[46] When Byron mentioned the war in Greece to his friends in Britain, the London Greek Committee wrote to the poet, inviting him to become its representative in Greece. He readily accepted. In transnational war volunteering, as in so many other things in life, timing is everything.

There are a few points of similarity between Byron's story and that of his longtime admirer, the renowned Polish poet Adam Mickiewicz. In 1855, during the Crimean War (1853–1856), Mickiewicz traveled to the Ottoman Empire to lend his support for the establishment of a Polish legion that would fight against tsarist Russia. Poland had been partitioned between its powerful neighbors—Russia, Prussia, and the Hapsburg Empire—in the late eighteenth century. A Polish uprising against Russia in 1830–1831 was crushed by the forces of Tsar Nicholas I. One of the hubs that kept the aspiration for Polish independence alive was the community of Polish exiles who lived in Paris. Mickiewicz formed part of that community and often found himself mediating between its different factions. In the early 1850s his personal circumstances changed. In 1852 he lost his job at the prestigious Collège de France and had to settle for a lower salary at a Parisian library. In 1854 his wife, Celina, became very ill. She died in March 1855. Although their marriage was seldom a happy one, the months-long ordeal of Celia's illness left Mickiewicz shaken and weary. When the leadership of the monarchist Polish exiles in Paris, with their headquarters at the Hôtel Lambert, encouraged the 56-year-old poet to travel to Turkey and lend his prestige to the establishment of a Polish legion he agreed. An acquaintance who was with him during his preparations for the journey in August noted that "Mickiewicz felt alive again." Once he and his companions were on their way, the poet seemed to be feeling "increasingly free, lively and hardy."[47]

Timing also played an important role in the decision of Henry Noel Brailsford, then a budding journalist from Glasgow, to volunteer to fight for the Greek cause against the Ottoman Empire in 1897. In December 1896 the 23-year-old Brailsford asked Jane Malloch to marry him, but she refused. A few months later he announced that he had made up his mind to go out to Greece as a volunteer if war was declared against Turkey. Once he learned of the formation of the Philhellenic Legion, he telegraphed offering his services to the Greek cause. Brailsford expressed his willingness to pay his own expenses. The war in Greece appealed to his romantic aspirations. According to his biographer,

> By committing his life to a noble cause, he would be released from absorption in his own subjective nature, freed from the tedium of the *Scots Pictorial* and his personal tribulations. At the same time the budding journalist in him recognized the adventure as "good copy," a long-sought opportunity to gain notoriety and prove himself worthy of a first-rate position.[48]

For Byron, Mickiewicz, and Brailsford, the decision to engage with a conflict abroad came at a time of changing circumstances or crises in their personal lives. In other cases the decision to volunteer was influenced by wider societal issues that affected large numbers of people. The impact of such push factors can easily be observed in the stories of numerous foreign volunteers in the Spanish Civil War.

Gustav Regler was a German antifascist writer. He voted against the incorporation of the Saar into Nazi Germany in the plebiscite of January 1935 and fled into exile immediately after the position he supported was defeated. Living in Paris initially, he moved to Moscow, where he was able to mingle with some of the Soviet Union's cultural and political elite. Lev Kamenev was one of his acquaintances. In the summer of 1936 Regler was following the news about the attempted coup in Spain with great interest, but he was also increasingly alarmed by the atmosphere of arrests and growing terror in Moscow. He applied to the Comintern for a recommendation for an exit visa in August, just as the show trial of Kamenev and other leading Bolsheviks was about to begin. "In Spain, I felt sure of it, I would breathe a different air," he wrote in his autobiography. "There, death was a protection against treachery and judges; one died at the hands of the enemy."[49] The Spanish Civil War offered a politically acceptable and face-saving way out for dozens of resident foreigners who wanted to leave the Soviet Union in 1936.[50]

For Jewish communists in Palestine, volunteering to fight in the Spanish Civil War also offered an honorable way out of a difficult predicament. In the 1930s they were part of a tiny party that was extremely unpopular among the country's Jewish population because of the anti-Zionist views it espoused. Communist Party members complained that they were being blacklisted by the Zionist-controlled labor unions and therefore struggled to find jobs. Another reason that pushed a number of communist Jews to leave Palestine was the Arab Revolt that began in spring 1936. The Palestine Communist Party (PCP) had pursued a policy of Arab–Jewish cooperation, but the Arab Revolt and the nationalist sentiments it aroused led to a de facto split in the party into separate Arab and Jewish sections. Members from the party's branch in Tel Aviv in particular were frustrated with the decision of the PCP's Central Committee to support the Arab nationalist camp.[51] Yet another problem that Jewish communists faced was the heavy-handed policy of the British Mandate authorities. Communist activity was illegal in Palestine. Party members who were arrested could, in some cases, be deported back to their country of birth.[52] For many, this would have meant deportation to Poland, where communist activity was repressed even more vigorously than in Palestine. PCP members, therefore, faced multiple pressures. Years later, in 1972, one former party member recalled: "We were caught between a rock and a hard place. . . . Against this background, there was something that attracted us in Spain—that we could go there without complications and with guns in our hands."[53]

Push factors of a different kind affected Spanish Civil War foreign volunteers from Britain and North America. In the 1930s, the years of the Great Depression, unemployment levels were very high on both sides of the Atlantic. The effects of poverty and occupational insecurity were especially potent for the urban working class from which the majority of volunteers from Britain, Canada, and the United States emerged. As one unemployed volunteer from southern Wales later explained, "I was an economic burden to my family and I thought I could be of more use in Spain."[54] In North America approximately 80 percent of the volunteers were immigrants themselves or had a parent who was born abroad, often a more economically vulnerable segment of society. The economic crisis drove many would-be volunteers into political activism, leading a considerable number to either support or join their local communist party.[55]

James Yates, the black American activist who decided to travel to Spain to fight both fascism and racism, had previously struggled to find a job in the United States because of the economic realities of the Depression as well as racial discrimination. A difficult atmosphere at home made him leave behind his wife and children in Chicago and move to New York. There the desperate social and economic conditions drove him into political activism.[56] One can only speculate whether Yates would have volunteered to fight in Spain had he been gainfully employed.

Volunteering for service in a foreign war, therefore, was (and is still) far more likely to occur in cases where the pull of ideology and adventure was supplemented by push factors that encouraged would-be volunteers to embark on such an endeavor. The combination of push and pull factors helps to explain why some supporters of foreign causes joined the fight while others, usually the majority, stayed behind. The relevance of such an explanation for contemporary foreign fighters in Syria and elsewhere is obvious. As Jessica Stern and J. M. Berger have recently noted, jihadi propaganda and awareness of conflicts abroad involving fellow Muslims "do not necessarily provide adequate motivation on their own merits. They offer outlets, either for social pressures in a fighter's native land or for his own internal struggles."[57] It is to these internal struggles and tribulations that we now turn our attention.

The Common Denominator

Stripped of the pull of ideology, the lure of adventure, and the specific circumstances that prompt them to leave—factors that in any case vary from one volunteer to the next—what remains is the common denominator. Individuals who volunteer to fight for a foreign cause invariably do so to find or to regain a sense of meaning. This search for meaning operates on two levels. The first is internal and relates to finding a purpose to which one can devote his or her efforts. The second is social and depends on how an individual positions him or herself in relation to others.

Man's Search for Meaning is the title of a book written by Viktor E. Frankl, a Viennese neurologist and psychiatrist. During the Holocaust, Frankl was a prisoner in Auschwitz, and his survival inspired him to develop a new psychotherapeutic approach that he called logotherapy (based on the Greek *logos*, or "reason"). Although it was developed in the 1940s and 1950s, Frankl's approach still provides some useful insights into the psychological mechanisms that lead to the decision of foreign volunteers to commit themselves to a cause.

Frankl believed that the search to find and give life meaning is a fundamental human driving force. Once a person has something to live for, she or he could contend with almost any circumstance. Of course, "meaning" changes from one individual to the next. A person's reason for living could also change over time. The task of logotherapy was to offer individuals an alternative point of view through which to see their personal predicaments, to help them identify a reason to live for. For Frankl, the object of the search need not necessarily be to gain pleasure or to avoid pain. Furthermore, "man's search for meaning and values may arouse inner tension rather than inner equilibrium."[58] But he saw mental health as being based on a certain degree of tension, "the tension between what one has already achieved and what one still ought to accomplish, or the gap between what one is and what one should become."[59] According to logotherapy, a person could discover meaning by doing a deed, by experiencing a value, or by suffering. He argued that "suffering ceases to be suffering in some way at the moment it finds a meaning, such as the meaning of a sacrifice."[60] Hence, people are prepared to suffer if it is for the right cause. Finally, Frankl believed that people had a degree of free choice and were "ultimately self-determining."[61]

Seen from a Franklian perspective, all foreign volunteers search for meaning, albeit in an intuitive manner. Emerging from a starting point of dissatisfaction with the present, the acuteness of which varies from person to person, the volunteers use the opportunity presented by the foreign conflict to improve their self-esteem. That individuals strive to maintain or enhance their self-esteem is one of the basic tenets of social psychology.[62] By committing themselves to a cause that they see as just, the volunteers also gain an opportunity to earn the respect of others.

The French sociologist, anthropologist, and philosopher Pierre Bourdieu also saw the search for meaning as a fundamental human action, though he emphasized the social aspects of this search for what he called the "justification for existing."[63] According to Bourdieu, individuals seek to fulfill meaningful roles in order to receive recognition from others. They struggle to emerge from indifference, depression, solitude, or insignificance in order to find a "feeling of counting for others, being *important* for them."[64] He argued that the social world alone can confer social importance, or what he called "symbolic capital." It is symbolic capital that rescues individuals from insignificance and gives their lives meaning.[65]

Foreign volunteers habitually expect to receive acknowledgment and recognition for their actions, although the level of recognition and form of acknowledgment may vary. Sometimes they expect to be recognized and acknowledged by friends, co-religionists, or party or ethnic group members in their home state. In other instances they may expect to be appreciated, perhaps even celebrated, by their wartime hosts. Their group of reference could be people they know or a broader like-minded imagined community. But the act of volunteering to fight in a foreign country has a social dimension to it that is intrinsically linked to the Franklian inner search for meaning. Bourdieu saw the state as "the central bank of symbolic capital" in the modern era. After all, the state can confer symbolic capital, social importance, and "reasons for living" on various occasions and levels.[66] Foreign volunteers, however, either cannot get the symbolic capital they expect in their home states or else feel that they might be able to obtain more of it by going to fight abroad.

Examples of how a search for meaning, both inner and in a social context, guided the decisions of foreign volunteers are plentiful. In 1855, having decided to contribute to the formation of a Polish force in Turkey to fight against Russia during the Crimean War, the poet Adam Mickiewicz hoped that his actions would inspire other Poles to join the cause: "Once [the Poles] see that I, with a gray head but with an ardent heart, am going where it directs me . . . then the young and the more able, since they are soldiers, would perhaps no longer dare to rot, beg, and not fulfill their sacred duty."[67] But alongside the social aspect of his voluntary activity, the act of traveling to Turkey in this manner also had an inner meaning for Mickiewicz. When he joined the Polish forces in their camp at Burgas by the Black Sea, he reportedly said, "Better to be a secretary in some regiment of Polish Cossacks" than "a chancellor of a French institute." To this he added, "even had [I] known that [I] would die from cholera somewhere in Turkey, [I] would still have gone."[68] This statement proved prophetic. Mickiewicz contracted cholera and died in Istanbul in November 1855.

Gustav Regler emerges from his autobiography as a restless man. Already during his frontline service in the German army in the First World War, "there were often things to be done that not everyone could be expected to undertake. I took them all on. It was a form of flight—running away by running forward."[69] Later in life his motivations continued to display a mixture of self-sacrifice, adventurism, and a deep-rooted empathy for the plight of others. For him, the decision to join the civil war in Spain was emblematic of his character. He observed that his "life has been governed by an inward compulsion: the important decisions have never been the outcome of long reflection but rather of a desire for simplification." He had always been inspired by "the thought that no man has the right to live self-indulgently in the presence of suffering, to take refuge in privacy from the problems of his time, or to hedge himself about with cynicism against his duty to his brother, his sick neighbour, his threatened friend."[70] Once in Madrid, Regler felt that, despite the practical difficulties he faced, he was where he wanted to be.

> I stood on the Gran Via without a *peseta* in my pocket and without knowing the language. I had no spare shirt and no letter of introduction to anyone. I was cast adrift in a way that would have induced panic in many people, but I felt happier than I had ever done in Paris or Moscow.[71]

During the Second World War Harold Livingston, from Haverhill, Massachusetts, had been an aircraft radio operator in the US Army Air Force. After being demobilized, he held a number of different jobs but was still waiting for what he called the "Big Break" to drop in his lap. According to his own account, his Jewish identity was peripheral: "My only real relationship to Jews and Jewishness was the vague perception that this made us, me, somehow different." When the Jewish baseball player Hank Greenberg hit fifty-eight home runs, Livingston "felt a strange tug of pride." He therefore summed up his approach to his Jewishness as the "Hank Greenberg Syndrome."[72]

Livingston had a vague awareness of what was going on in Palestine after the Second World War, finding news pertaining to the brewing conflict there "interesting, but of little personal significance beyond being part of the Hank Greenberg Syndrome." Nonetheless, when an acquaintance from his past sent him a letter, suggesting that Livingston join an "outfit being formed to fly munitions and refugees to Palestine," he accepted almost without hesitation.[73] Much like Regler, Livingston explained his motivations using terminology that combined ideals with a strong dose of adventurism and an attraction to exotic and "exciting places."[74]

> I was helping an oppressed people establish their own country. This is what I told myself, and it was basically true, but I also wanted action. The excitement. The Old Days. That it was a Jewish operation, to be sure, was much of the motivation. But I liked to believe that what really motivated me was that I, an American, would be helping those people fight for, and establish, a nation in the finest American tradition.[75]

Livingston found what he was looking for. "In wartime your entire existence is accelerated," he wrote in his memoir. Despite the difficult and tiring routine of flying supplies to the newly established State of Israel as well as within the war-torn country, and even though he had lost friends in the process, he "wanted to feel again that warm flush of achievement and self-esteem, strolling along a Tel Aviv boulevard knowing I was recognized as an American volunteer. I wanted that exquisite sense of acceptance and belonging."[76]

Aimen Dean's motivations for volunteering to fight in the war in Bosnia in the 1990s echo some of the sentiments raised in Livingston's account. Dean told BBC Radio 4 about the sense of purpose, empowerment, and camaraderie that service during the conflict gave him:

> It was the most eye-opening experience I ever had. I was a bookish nerd from Saudi Arabia just weeks ago and then suddenly I found myself prouncing [sic] up on the mountains of Bosnia holding an AK47, feeling a sense of immense empowerment. And the feel that I was participating in writing history rather than just watching history. And also at the same time being in the military training camps, receiving knowledge that I never thought in a thousand years I would be receiving about warfare and war tactics and military manoeuvres, and to be receiving it alongside people from many different nationalities with the one common factor among them that they were all Muslims.[77]

Roland Bartetzko, who was born in Würselen near Aachen in Germany in 1970, also fought in Bosnia-Herzegovina, though he joined a Bosnian-Croat unit. He had been a staff sergeant in the German army until 1992 and then entered into higher education. However, he left without graduating to pursue what he termed as his "calling." Looking back at his motivations, he described both moral reasons for enlisting as well as personal considerations that were not directly related to the war in the Balkans. On the one hand, he believed that the Croats were "fighting for a just cause, otherwise I wouldn't have joined them." On the other hand, he

> read a lot of stuff, mostly philosophical texts (Maeterlinck, Thomas Mann, all this "fin du siècle' stuff like Baudelaire, but also Nietzsche, whom I carried along with me during the war), before I joined the war in Bosnia, and thought that war would be an experience that I needed to achieve a higher level of consciousness; one could say that fighting a war was a milestone for my road (or quest) to self-fulfillment.[78]

In the late 1990s Bartetzko went on to join the Kosovo Liberation Army, an organization that fought for the independence of Kosovar Albanians from Serbia. Once again, his motivations combined ideals and self-fulfillment. He was attracted to the idea of participating in "a guerrilla war," something he had trained for while in the German armed forces, but never got a chance to experience. And then there were moral considerations:

> After I saw what the Serbs where doing in Bosnia, I couldn't sit and wait that somebody else [will do] something to help the Kosovar population. . . . The cause was absolutely and undeniably just. Sometimes I had my doubts [when] in Bosnia [whether] I was fighting for the right side.[79]

A similar mix of motivations, underlined by a search for purpose, can be observed among Western volunteers who have joined Kurdish forces in their fight against the Islamic State in Syria.[80] We have already seen how the Canadian John Gallagher saw this conflict as part of a broader struggle against theocracy. A former soldier in the Canadian armed forces, he had graduated with a bachelor's degree in political science and gender studies in 2012 and went on to complete his master's degree in 2014. Gallagher considered continuing his studies at PhD level, but in the end decided not to do so. Before leaving Canada he made deliveries so that he could pay the bills, set up an online editing business to cater for university students, and dedicated himself to his own writing. Valerie Carder, his mother, later spoke of "the frustration he felt in general about not being where he wanted to be in life, not effecting change in the way that he wanted."[81] Gallagher traveled to join Kurdish forces in spring 2015. In August of that year he posted a half-humorous update online, in which he announced:

> I have to admit I'm having the time of my life out here. If you ever
> have the chance, I strongly recommend fighting terrorists as a great
> way to get yourself out of your various existential funks. True, the
> diarrhea may kill you if the mortars and car bombs don't, but you'll
> die knowing that you pissed off a lot of . . . hippies.[82]

After Gallagher's death, his mother said that he had not expected to be celebrated as a hero. However, he "wanted to make a difference; he wanted to do something important with his life; he wanted people to know his name; he wanted people to read his work."[83]

As the examples of Mickiewicz, Regler, Livingston, Dean, Bartetzko, and Gallagher illustrate, the act of volunteering to fight for a foreign cause gave these individuals a sense of purpose and self-fulfillment. Each one of them faced different predicaments at home before volunteering. It could be exile, as in the case of Regler, or boredom, as in the case of Livingston. But for all these individuals something was missing. Indeed, though the reasons vary, all foreign volunteers share a sense of what Frankl called a "gap between what one is and what one should become."[84] These examples also illustrate how the act of volunteering provided opportunities not only for supporting an ostensibly just cause but also for experiencing excitement and wartime masculine camaraderie.

Gender Wars

Nearly all the volunteers that have been discussed until now have been men. This is hardly surprising when we consider that the overwhelming majority of foreign war volunteers in the modern and contemporary eras have been men. Out of approximately 2,500 British volunteers who offered their services in various capacities to the Spanish Republic in the late 1930s, only about 70 were women.[85] A survey that was conducted among foreign war volunteers who remained in Israel shortly after the 1948 war had ended found that 90 percent of them were men.[86] In 2009 the Association of Foreign Volunteers of the Croatian Homeland War compiled a list of more than 450 foreigners who had enlisted in Croatian units during the Yugoslav Wars. The association's roll of honor included only one woman.[87] Gender identities and especially the desire to affirm one's masculinity go a long way toward explaining why certain individuals were, and are still, attracted to the prospect of taking up arms in a foreign conflict.

Lord Byron had a notoriously complicated relationship with women. While sailing to Greece, among the all-male passengers and crew of the brig *Hercules,* he gave voice to some views we would now call anti-feminist. According to fellow traveler James Hamilton Browne, Byron "often contended in favour of the Oriental custom of secluding females, and teaching them only a few pleasing accomplishments." Furthermore, he believed that "the Greeks were sensible people in not allowing their daughters to be instructed in writing."[88] One scholar has gone as far as to argue that Byron's decision to volunteer his services for the Greek cause in 1823 was partly motivated by a desire to relive some of the homosexual encounters he had had when he last visited Greece, more than a decade earlier.[89]

The young journalist Henry Noel Brailsford pursued a different kind of masculinity when he volunteered to fight for Greece in 1897. Jane, the woman he wanted to marry, urged him to go to Greece, telling him that she would not have hesitated to do so if she were a man. Meanwhile Brailsford "worried about whether he would learn to handle a rifle properly, never having fired one before."[90] When he returned from Greece he published *The Broom of the War-God,* a memoir disguised as a novel. In it, the protagonist, Graham, helps carry an injured comrade: "He envied his rude companions; he would have given all the three years of life spent in studies and class-room for the muscles of a mason or carpenter." When a bullet passed through his sleeve, grazing his arm's skin, "he felt an insane delight in this toy danger."[91]

As we have already seen, Peter Kemp, one of the few British volunteers who fought for Franco, listed a number of motivations for going to Spain in 1936. One of these related to his strained relationship with his father:

"Manliness" and a serious attitude to work were the qualities he [the father] looked for, and sadly failed to find, in his younger son; he considered me feckless, indolent and feeble. One of the reasons—by no means the only one—why I had come to Spain was to make him change his opinion.[92]

Kemp was not unique in his desire to live up to what he saw as familial expectations. Volunteers in different conflicts also saw their service abroad in terms of fulfilling their families' military traditions.[93]

For the aircrew member Harold Livingston, war was, among other things, an opportunity to meet women. When he was interviewed for the documentary film *Above and Beyond* (2015), he recalled that in 1948 "we were young, good looking, had hair, we had a lot of testosterone and we didn't want to waste it."[94] Che Guevara, who had long considered himself "a man of action," was much more austere.[95] In August 1967, a few weeks before his death, he tried to galvanize his guerrilla band in Bolivia by telling them that "this type of struggle gives us the opportunity to become revolutionaries, but also allows us to prove we are men. Those who are unable to reach either of these stages must say so and leave the struggle."[96] As these examples illustrate, the search for masculinity formed part of a broader search for meaning for men in different periods and coming from different cultures.

Statistically, women were far more likely to volunteer in non-military contexts. For instance, women constituted approximately 47 percent of the wave of more than 7,000 volunteers from several countries who went to Israel in the summer and fall of 1967 to work in agriculture, reconstruction, and other temporary jobs following the Six-Day War.[97] In the 1980s, when several hundred leftist activists from the United States, Britain, and other countries traveled to Nicaragua to join volunteer work brigades that had been set up by Nicaragua's Sandinista government, female volunteers were numerous.[98]

When women have enlisted as foreign volunteers in times of war, they have usually been assigned to traditionally "feminine" roles. One of the most prevalent of these was nursing. Patience Darton from southern England was 25 when she contacted the Spanish Medical Aid Committee in London and offered her services to the Spanish Republic in early 1937. Qualified in nursing and midwifery, she held left-leaning views but was not politically active. When the civil war broke out, she was impressed by the participation of ordinary people in Spain in the struggle. Years later she recalled thinking: "I'm a nurse, I can do something too, because mostly women couldn't." Echoing the search for purpose shared by her male counterparts, she stated: "I didn't know how to get into it, but I knew perfectly well, this was the place, there [in Spain] they were doing something, and I was lucky, I could do something too."[99]

Female foreign volunteers have also been used in clerical capacities. Lily Myerson from London went to Israel in 1948 to find her husband and brother, who had already volunteered for military service. She too enlisted and served first as a secretary at the army headquarters before transferring to the air force. Her services were needed in the air force because the majority of its personnel came from Britain, Canada, South Africa, and the United States and the lingua franca was English.[100]

Some female foreign volunteers who went to Israel in 1948 were assigned to training duties. Anita Koifman was a Second World War veteran who had served in Britain's RAF Fighter Command headquarters. Coming from a Zionist-Jewish family, she was engaged in fund-raising for the Jewish cause in Palestine while still in Britain. Following the Israeli declaration of independence, she wanted to volunteer but was discouraged from doing so by those around her. She was told, "You'll be just another woman there, and they only need nurses anyway." Once she reached Israel, air force officials wanted to make use of her experience in radar. Thus, Koifman ended up helping to design the Israeli air force's operations room, drawing on her knowledge from the Second World War. She also taught plotting and trained new Israeli recruits to become radar operators.[101]

In some instances, women were used for recruitment purposes. In April 1987, for instance, as part of its campaign to recruit foreign volunteers for the Afghan conflict against the Soviet Union, the magazine *Al-Jihad* featured what was then an unusual appeal from a woman who wrote:

> I urge you, oh young men, to undertake Jihad and martyr yourselves beside your Afghan brothers who are fighting against oppression. I only wish I could give my life and my spirit as a gift to this pure land as a martyr. But I am a girl and not able to do anything.[102]

In comparison to the roles that female foreign volunteers have traditionally been assigned in the past, there is something novel about the approach of the Islamic State toward the women it recruits from abroad. As we have already seen, the organization encouraged Sunni Muslims from all over the world to immigrate to the territories it held in Iraq and Syria. This included young women, sometimes still in their teens. Aqsa Mahmood, a 20-year-old former student from Glasgow, traveled to Syria in 2014 and married one of the organization's fighters. She also played an active role in recruiting other Western women through social media. Islamic State propaganda makes it clear that, once they arrive, foreign women should expect to become wives to jihadists and bearers of children for the Islamic State. As Bint Nur, the wife of a British fighter in Syria, wrote in 2014, "women build the men and men build the Umma."[103] With their numbers reaching several hundred by 2015, some foreign women have demanded to take on a more combative role and have been observed in frontline capacities. Meanwhile a manifesto issued by the Islamic State sought to make clear that women are only permitted to abandon domestic roles for fighting "if the enemy is attacking her country and the men are not enough to protect it, and the imams give a fatwa for it."[104]

If we look back at the historical record, there have been exceptional cases where foreign women have taken on frontline combat roles. There is considerable legend surrounding Anita Garibaldi, the Brazilian-born wife of Giuseppe Garibaldi. She reportedly fought alongside her husband in a naval engagement in Brazil in 1839, taunting those sailors who hid below deck and refused to take up arms. Anita was also in Rome in 1849 and, after the defeat of the short-lived republic, she dressed as a man and accompanied Giuseppe and his troops on their long march north. When their force was attacked by Austrian troops near San Marino, Anita tried to stop the men who had begun to escape, though with little success. Pregnant and ill as a result of the arduous journey, she died later that summer.[105] But perhaps the best-known example of a frontline female foreign volunteer is Flora Sandes, the British woman who joined the Serbian army during the First World War. Sandes begins her autobiography by declaring:

> When a very small child I used to pray every night that I might wake
> up in the morning and find myself a boy. . . . Many years afterwards,
> when I had long realized that if you have the misfortune to be born

a woman it is better to make the best of a bad job, and not try to be a bad imitation of a man, I was suddenly pitchforked into the Serbian Army, and for seven years lived practically a man's life.[106]

Already well-traveled and familiar with "camping out and roughing it," Sandes joined a volunteer St. John ambulance unit as a Red Cross nurse and was posted to Serbia in 1914.[107] When the regiment to which she was attached began to retreat toward Albania in late 1915, "where there were no roads, and we could not take . . . ambulances to carry the sick, I took the Red Cross off my arm and said, very well, I would join the 2nd Infantry Regiment as a private."[108] Her Serbian hosts agreed to enlist her, being more impressed by the fact that she was British than by her being a woman:

> The soldiers in the Ambulance seemed to take it for granted that anyone who could ride and shoot, and I could do both, would be a soldier, in such a crisis. To their minds there was nothing particularly strange about a woman joining up, there had occasionally been Serbian peasant girls in the army. . . . The only thing that distinguished me particularly, and made them treat me with so much affection and respect, was the fact that I, an Englishwoman, was willing to rough it with them, and to fight for Serbia.[109]

Once she began her military service, Sandes was able to find the same sense of camaraderie and self-fulfillment that so many male foreign volunteers have described. Officers and men called her " 'Brother'—the usual term of address among Serbs." She recalled how

> my tough and hardy comrades, most of them young veterans of two previous wars, taught me how to be a Serbian soldier. . . . For us, our own company was a hub of the Universe, particularly our own platoon. . . . I seemed to take to soldiering like a duck to water.[110]

Sandes was able to fight for Serbia at a time when combat roles were off limits to women in her native Britain. The ability of female volunteers to participate in frontline military service is therefore dependent on the level of openness toward women exhibited by the military organization they join. The early days of the Spanish Civil War in the summer of 1936 provided opportunities for women, both local and foreign, to take up arms. Fanny

Schoonheyt from Rotterdam in the Netherlands was living in Barcelona when the conflict erupted. She joined one of the hastily organized antifascist militias and saw action on the Aragón front, where her technical prowess earned her the nickname "the queen of the machinegun." She later served in the Republican army but did not join the International Brigades.[111] Another foreign woman who was in Spain when the war began was Felicia Browne, a British sculptress who was killed at the front on 25 August. Her obituary in the *Daily Worker* described her as having "a fine courageous and adventurous spirit."[112] Before leaving for the front, Browne told a British reporter: "I am a member of the London communists and can fight as well as any man."[113]

However, the incorporation of women—both Spanish and foreign—in fighting units was phased out once the military forces of the republic became more organized. Two Jewish women from Palestine who initially trained with the men of the Polish Dabrowski (also known as "Dombrowski") Battalion were recalled to the International Brigades headquarters at Albacete in the autumn of 1936. Much to their dissatisfaction, they were reassigned to clerical positions.[114] The Republican leadership decided not to use female fighters for a number of reasons. Having women on the frontline risked offending public opinion. There was also the fear—sometimes realized—of the fate that might befall a female fighter once captured by the enemy. And then there was the argument, often repeated in debates about the incorporation of women into fighting roles in other countries throughout the twentieth century, that a wounded female soldier would put the lives of her male comrades at risk because they would feel more compelled to rescue her.[115] From that point onward, the command of the International Brigades did not encourage foreign women to come as fighters, and communist parties abroad did not actively recruit them.[116] Women who did join the International Brigades served as nurses, ambulance or truck drivers, clerks and in various other supporting roles. To use a phrase coined by feminist scholars, the foreign women who participated in the Spanish Civil War rarely got a chance to defy "male civilization."[117]

One military organization where the incorporation of women—mainly local but in some instances also foreign—in fighting roles has been much more prevalent is the PKK and its affiliate in Syria, the YPG. We have already seen how Andrea Wolf, a leftist feminist from Germany, joined the PKK in the 1990s and died in its ranks. In the civil war in Syria some 40 percent of the personnel of Kurdish militia are women.[118] Gill Rosenberg, a 31-year-old Canadian-born citizen of Israel, received international media attention in 2014 when she became the first foreign woman to join Kurdish forces in Syria. Upon returning to Israel she said that the atrocities committed by the Islamic State against Yezidi women prompted her to join the Kurds. The fact that their units include both male and female combatants made her decision easier.[119] This suggests that organizations such as the PKK and the YPG, which include the emancipation of women as one of their goals, can provide opportunities not only for foreign men to affirm their masculinity but also for women to challenge socially constructed gender roles.

* 4 *

Thoughts of Home

A Typology of Volunteer–State Relations

When examining the motivations of foreign volunteers, historians, analysts, journalists, and sometimes the volunteers themselves have often sought to divide their subjects into categories. For instance, according to Robin Winks, volunteers from Canada—or British North America—who fought in the American Civil War between 1861 and 1865, did so for the following reasons: "Some undoubtedly fought for the love of adventure, some fought as a private crusade against slavery, some fought because they ordinarily had enlisted for the bounty and found that they liked the life, some fought because they had nothing else to do."[1] Jason Gurney, then a 26-year-old sculptor from Chelsea, London, joined the International Brigades in 1936 and fought in the Spanish Civil War. He retrospectively described his battalion as a "mixed bag, drawn from all classes of society." It included "pure idealists, political opportunists, doctrine Marxists, adventurers and plain rogues, in varying proportions."[2] When assessing the motivations of Al-Qaeda recruits in the early 2000s, psychological operations officer Colonel John Venhaus of the US Army distinguished between four categories of what he called "seekers": revenge seekers, who perceive themselves as victims in society; status seekers, who want to stand out from the masses but see a world that does not understand or appreciate them; identity seekers, who wish to bolster their self-esteem by assimilating into a defining organization or group; and thrill seekers, who are filled with energy and drive, and want to prove their manhood.[3]

The reporter Jennifer Percy, who interviewed Western volunteers fighting alongside the Kurds against Islamic State militants in 2015, explained why they were there: "Some were escaping life back home. Others were old soldiers, trying to fill a void. A few just had delusions of grandeur. They came for the feeling of solidarity, or adventurism, or they came to fulfill a childhood fantasy, to act out some violent adolescent emotion."[4]

The subdivision and classification of any complex phenomenon into distinct types is a tool that provides clarity. In a way, classifying is understanding. But there are also risks in using categories—namely, oversimplification and misrepresentation. As we have seen in the previous chapters, individual volunteers routinely had more than one reason for wanting to fight in a foreign conflict. Ideological or political motivations were often mixed with personal reasons. The pursuit of a better status, a clearer sense of identity, or improved self-esteem did not preclude a simultaneous search for excitement and adventure. How, then, can we distinguish between different types of volunteers if motivations on their own do not give us a clear answer?

This chapter examines the ways in which volunteers have positioned themselves in relation to their home state and, to a lesser extent, their wartime hosts. It suggests that there are four fairly distinct categories of volunteer–state relations that provide a more helpful method of understanding the phenomenon than divisions based on motivations per se: self-appointed ambassadors, who see themselves as fulfilling a task that should have been carried out by their own government; diaspora volunteers, whose willingness to enlist is tied to a military crisis in their country of heritage; cross-border volunteers, who share national or ethnic ties with a neighboring group engaged in conflict; and substitute-conflict volunteers, who see service in a conflict abroad as a precursor to fighting the regime in their home state. Of course, the world that the historian sees is rarely divided into neat categories, and there are some overlaps in the classification presented here. However, the assessment of the volunteers' attitudes toward their home state in particular offers a new way of understanding which of them pose a threat to their country's national security and which do not.

Self-Appointed Ambassadors

Volunteers belonging to the self-appointed ambassador category see themselves as patriots of their home state. These are individuals who insert themselves in conflicts in which their home state has opted to remain on the sidelines. Such volunteers tend to perceive their governments as unnecessarily or unjustly passive and hesitant, while they view themselves as representing the true values of their nation.

The nineteenth-century Concert of Europe, with its emphasis on entrenching "legitimate" monarchist rule, smothering revolutionary movements, and resisting as far as possible any alterations to the continent's borders, provided the backdrop against which a number of self-appointed ambassadors offered their military services. Historian Davide Rodogno has shown how, at various junctures, the lack of government-sanctioned military intervention by the great powers against the Ottoman Empire created a vacuum that individual volunteers sought or felt obliged to fill. The London Greek Committee, which was founded in 1823 and eventually used the services of Lord Byron as its emissary, promoted the causes of liberty, free trade, and a republican government for Greece. The committee also became a locus for opposition to the politics of the Holy Alliance between Austria, Prussia, and Russia, which spearheaded the effort to preserve the post-Napoleonic status quo in Europe. The philhellenes in London and Paris, and the volunteers who traveled to Greece, not only kept the question of intervention on behalf of the Greeks alive until it materialized in 1827 but also turned the Greek issue into one that could bring policy makers moral capital.[5]

Similarly, volunteers from Italy, Hungary, France, and mainland Greece who landed in Ottoman-controlled Crete in late 1866 wanted to sustain the local revolt until the spring to force the great powers to intervene in the conflict on behalf of the island's Christian population.[6] This type of volunteering occurred in other parts of Europe as well. Historian Gilles Pécout coined the phrase the "diplomacy of the people" to denote the decision of foreigners to voluntarily take part in the wars of the Italian Risorgimento. He contrasted this form of transnational grassroots activism with the "diplomacy of the cabinets," which was perceived by the volunteers as stifling the spirit of liberty and national emancipation.[7]

The Russian general Mikhail Grigorevich Cherniaev was a nineteenth-century self-appointed ambassador who was propelled into action by a different set of ideological motivations. Hailed as the "Lion of Tashkent" after leading the Russian force that captured the city in 1865, Cherniaev sought to champion the cause of Pan-Slavism, the perception that all the Slavic peoples should unite under Russian leadership. Fired by personal ambition, the half-retired general tried to organize a military expedition to Herzegovina in September 1875 to assist the local Christian population in its rebellion against the Ottoman Empire. Tsar Alexander II was alarmed by this prospect and placed Cherniaev under surveillance. Meanwhile, uprisings against the Ottomans had also spread to Bulgaria, and tensions throughout the Balkans were running high. Despite the tsar's objection, Cherniaev managed to enlist the support of the Slav Committee in St. Petersburg and left Russia surreptitiously in April 1876. He traveled to Serbia, where he accepted command of the army of the Morava and was soon joined by a few thousand fellow Russian volunteers. The general saw Serbia as the key to the liberation of the Balkans, but he also sought to become Serbia's military dictator and to reshape its institutions. By the summer of 1876, Serbia and its foreign volunteers were at war with Turkey.[8] As we shall see later, Cherniaev and his Serbian allies were defeated. However, within a few months the initially reluctant Alexander II declared war on Turkey, ordering his army into the Balkans.

Another volunteer who conflated his personal ambitions and frustrations with his opinions on the international standing of his nation was Colonel Georges Henri Anne-Marie Victor, Comte de Villebois-Mareuil. A decorated officer, he left the French army in December 1895. He did so because the war ministry planned to promote him to the rank of general and to assign him to serve in metropolitan France rather than posting him to command the 1st Regiment of the Foreign Legion in Madagascar, as he had wanted. When he offered his services to the Boers of southern Africa in their struggle against the British Empire four years later, Villebois-Mareuil explained that he had ended his military career "because I found out that France's international policies were harming increasingly the vitality of the Army."[9] His concern for the supposed degeneration of France prompted him, during those four years, to establish the Union des Sociétés Régimentaires, a nationwide organization of ex-soldiers, and to cofound the political movement Action Française, which later gained notoriety for its activities in the lead-up to and during the Second World War. Villebois-Mareuil was appalled by the internal crises that rocked French society at the time, the most famous of these being the Dreyfus affair.[10] He was also outraged by France's weakness on the international stage as the Fashoda incident of 1898 had illustrated. This standoff between French and British forces in Sudan offended Villebois-Mareuil, who was a great believer in the honor of the army. He saw the small French force, which was compelled by the government in Paris to retreat so that war with Britain could be averted, as symbolizing everything that was wrong with the Third Republic.[11]

Villebois-Mareuil arrived in Pretoria in November 1899. His impression of the Boers was mixed. On the one hand, he was impressed by their spirit of freedom, but on the other, he was very critical of their lack of military organization. In his diary he lamented the fact that the French government was not doing more to help: "The Boers still rely on aid from France, because they are living on their illusions of two hundred years ago. But we have got beyond that."[12] The colonel therefore saw himself as retrieving France's lost glory while also fulfilling his own destiny. On 19 March 1900 Villebois-Mareuil received the rank of general and was given command of a small international unit, which consisted mainly of French and Dutch volunteers. As one of his admirers back in France, the author and diplomat Eugène-Melchior de Vogüé, wrote shortly afterward:

> He had but one object in view, one pre-eminent thought which appears many times between the lines of his diary. This thought was no longer for the Boers, but for France and her army. He had sworn to leave in the depths of Africa an imperishable recollection of French bravery; he had resolved to show all—friends or foes—how the soldiers of his nation could die.[13]

And die he did. On 5 April 1900 Villebois-Mareuil led a small body of foreign volunteers on a reconnaissance mission near Boshof. When they were surrounded by a much larger British force, Villebois-Mareuil refused to surrender and was killed after putting up a stubborn defense.[14]

Many of the British volunteers who took part in the Spanish Civil War fall into the self-appointed ambassador category. For example, Charlotte Haldane, who was sent by the Communist Party of Great Britain to Paris in 1937 to assist with the transportation to Spain of British volunteers for the International Brigades, explained in her memoir that

> the fact that I would be aiding and abetting a transaction declared illegal by the British Government did not worry me at all. I was wholly on the side of the International Brigade and opposed to the Chamberlain Government's policy on Spain, disgusted by its apparent fraternisation with the German Nazis and the Italian Fascists.[15]

Louis Hearst, who arrived in Spain from London in September 1936, was angered by Britain's policy of neutrality and non-intervention in the war in Spain. As he later recalled,

> When England took the reins of "Non-intervention" I could hold back no longer. I put everything else aside, and decided that I could not identify myself even passively (being a Socialist both by personal conviction and by family tradition) with this political farce; I decided to go to Spain to defend democracy with deeds.[16]

In other words, both Haldane and Hearst not only sought to help the Spanish Republic but also saw their voluntary commitment as a form of protest against Britain's foreign policy.

The Winter War between Finland and the Soviet Union provided another opportunity for volunteers who sought to represent what they perceived as the true character of their home state. The Soviet Union launched its attack against Finland in late November 1939. The Finnish cause was very popular among large sectors of the public in neighboring Sweden. Rickard Sandler, Sweden's foreign minister, suggested that his country should take concrete military action in support of the Finns, but his proposal did not gain the government's support and he resigned. Wary of antagonizing Stalin, Stockholm sought to preserve its neutrality, especially in the unpredictable circumstances created by the outbreak of the Second World War. Geoffrey Cox, a British journalist who passed through Sweden on his way to Finland in December 1939, observed that the Swedes were "torn by anxiety about the Finnish war, by a blend of fear that the war would spread to Sweden and shame that they were not already fighting in it."[17]

There were various efforts to recruit volunteers for Finland. While the Swedish government could not give its public support, it is clear that these efforts received tacit backing from above. Cox reported seeing full-page advertisements in the papers, exclaiming "Finland's fate is yours."[18] Pro-Finnish propaganda material appealed to the patriotism of prospective volunteers by stating: "Now the world knows what it means to be a Finn. Show the world what it means to be a Swede." The shared Nordic identity of the two countries was also invoked. According to one poster, "Finland's fight is the fight of the Nordic countries, which is the fight of the Western world." The struggle was also presented in terms of defending democracy and freedom. The Swedish author and historian Vilhelm Moberg published a letter to Sweden's farmers warning that "the Red Army is on the march toward Scandinavia." He added, "This is the time for Swedish farmers to show the world the value they place on the culture they have nurtured for a thousand years."[19] As we shall see in Chapter 5, some 8,000 Swedish volunteers answered the call and took part in the Winter War between Finland and the Soviet Union before it ended in March 1940.

While most of the foreign volunteers who fought for Israel in the war of 1948 do not fall into the self-appointed ambassador category, a few did. This was the case with Gordon Levett, a former RAF pilot from London. Although he was not Jewish, Levett was sympathetic toward the Zionist cause and was opposed to the way in which Britain was handling its withdrawal from Palestine. In his memoir, he explained that he could not fathom why Britain was not actively aiding the establishment of a Jewish state "after the horror of the gas chambers of Europe had been fully exposed."[20] Like many supporters of Zionism at the time, he perceived the policy of Britain's Labour government as being pro-Arab. Levett also believed the statements made by some of the Arab leaders in 1948, claiming that they would drive the Jews into the sea. "To this day I wonder what [Prime Minister Clement] Attlee and [Foreign Secretary Ernest] Bevin and their Cabinet colleagues would have said when the last Israeli Jew and Jewess had drowned off [the coast of] Tel Aviv."[21] Reflecting further on his motivations, Levett told the *New York Times* in 1998: "I felt anger and shame. . . . There comes a time when morality overrides politics."[22] Although the Zionists initially suspected that Levett was a British spy, he went on to fly several supply and combat missions for the IDF.

The decision of self-appointed ambassadors to commit themselves to a foreign cause could be an act of defiance against the policy of their home state but also a response to the inaction of the international community at large.

This was the case with Ivan Farina, an Irish volunteer who joined Croatian forces fighting against the Serbs in northern Bosnia in 1992. Like so many other foreign volunteers before and after him, he was motivated by a combination of personal and political reasons. He believed that the emerging independent nations in the former Yugoslavia had a right to be free but also felt dissatisfied with his life in London when the war broke out. However, for our present purpose it is important to highlight how he positioned himself with regard to Western policy toward the conflict in the Balkans. As he would later recall,

> the lack of international support, especially in the face of over-whelming Serb aggression bothered me a lot. When the existence of concentration camps came to light in Bosnia all doubts were dispelled and I concentrated on getting over to fight. As soon as funds permitted I left for former Yugoslavia.[23]

Finally, wanting to put forward a positive image instead of the policy promoted by one's government is also present among non-military volunteers. For instance, citizens of the United States who resented President Ronald Reagan's policy of supporting the Nicaraguan Contras in the 1980s traveled to Nicaragua and carried out voluntary agricultural work for the left-wing Sandinista government.[24] Sox Sperry, one of the volunteers, explained in a letter to friends and family back home:

> I decided to go to Nicaragua to participate in the harvest in response to the Nicaraguan government's request for assistance from North American friends. This action struck me as an active and nonviolent way to oppose the Reagan government's warmaking in the region, to support the Nicaraguan people's right to determine their own future, and to gain first-hand knowledge of the situation in Nicaragua.[25]

Self-appointed ambassadors, therefore, engage in a form of "diplomacy of the people." They seek not only to assist one side in a conflict abroad but also to influence, or at the very least protest against, the policy of their government back home. British intervention in Greece in 1821, Swedish intervention on Finland's side in late 1939, or Western intervention in the war in former Yugoslavia in 1992 would have deprived the individuals discussed here from a major reason for volunteering.

Diaspora Volunteers

Different scholars have defined the term "diaspora" in different ways.[26] Initially referring to the historical dispersion of Jews beyond the Land of Israel, today the term is often used in a much broader sense, applying to any group of people living beyond their traditional homeland. The problem with this broad demarcation of diaspora is that it may include people who do not see themselves as being part of such a group. Therefore, for our purpose, use will be made of the definition put forward by the scholar of political violence Jolle Demmers. For her, diasporas are "collectives of individuals who identify themselves, and are identified by others as part of an imagined community that has been dispersed (either forced or voluntary) from its original homeland to two or more host-countries and that is committed to the maintenance or restoration of this homeland."[27] Included in this definition are members of ethnic minority communities that are well established in their country of residence but also have a country of heritage to which they feel an affinity for either historical, religious, cultural, or familial reasons.

Diaspora volunteers can be defined as those who become involved in a conflict because of such ties to the country of heritage. In times of conflict, appeals from the beleaguered historical "homeland," media portrayal of suffering there, and sometimes even unsubstantiated rumors have been known to resonate with diaspora communities. While people might not always identify primarily as members of a diaspora community, specific events can trigger such identification. A military crisis involving co-ethnics or co-religionists abroad, for instance, can raise the salience of those identity categories that mark out a diaspora member.[28]

Diaspora activism in support of the country of heritage has often been restricted to political lobbying and fund-raising in the state of residence. In some cases it has also included the shipment of arms but stopped short of dispatching foreign volunteers to take part in the conflict. For instance, in the 1970s, during the "troubles" in Northern Ireland, Irish Americans raised hundreds of thousands of dollars per year for the Irish Republican Army. Some twenty people were also convicted in the United States on charges relating to gunrunning for the militant Catholic organization during the same period.[29] However, this support did not translate into a movement of foreign volunteers from the United States who were prepared to actively fight against British troops in Northern Ireland. Our focus here, therefore, will be on cases where diaspora members crossed the military participation threshold: the Greek diaspora in the 1820s, the Jewish diaspora in 1948, and the Armenian, Croatian, and Kurdish diasporas in more recent decades. As we have already seen, the number of those who volunteered their military services always constituted a small minority within a larger pool of potential diaspora supporters.

In 1821 Greek diaspora communities played a key role in launching the revolt against the Ottoman Empire. For instance, Alexander (Alexandros) Ypsilantis, who had served in the Russian army during the Napoleonic Wars, led the initial, unsuccessful uprising against the Ottomans in Moldavia. Meanwhile his brother, Demetrios, also a former officer in the Russian army, was appointed by the secret Greek nationalist organization, the Friendly Society, to lead the revolt in the Peloponnese. Initial successes in southern Greece created a wave of enthusiasm among Greek diaspora communities from western Europe to Egypt. According to historian William St. Clair, "As soon as they heard of the Greek 'victories' in the Peloponnese, hundreds of Greeks studying in European universities or working in merchant houses made their way to the sea and embarked for the homeland which few of them had ever seen."[30] Demetrios Ypsilantis wanted to create a European-style army. Some of the other diaspora volunteers joined a regiment that he sought to establish under the command of a Frenchman, Colonel Baleste. However, Ypsilantis found it difficult to communicate with the local Greeks and was unable to convince them to join his endeavor. He soon ran out of money, and Baleste left for Crete, where he was killed in battle.[31] By 1824 relations between the so-called Europeanized political elite, which included diaspora dignitaries who returned to their historical homeland, and local Greek warlords were openly hostile. Internecine violence followed, and it was only the landing of Egyptian forces in the Peloponnese under Ibrahim Pasha in 1825 that shifted Greek attention back to the conflict against the Ottomans.[32] Although Demetrios Ypsilantis eventually achieved the rank of field marshal in 1828, the attempt to fuse diaspora military know-how with local operational practices in Greece in the 1820s was, compared to later cases, particularly unsuccessful.

When the newly established State of Israel fought its first major war against the neighboring Arab countries in 1948, thousands of Jews from around the world volunteered their services. For some of these volunteers, their Jewish identity was central to who they were, even if they were not particularly observant. For the 20-year-old Vidal Sassoon, volunteering was about being Jewish rather than about Judaism. Raised in poverty, Sassoon spent seven years of his childhood in London's Spanish and Portuguese Jewish Orphanage because his mother, who had been evicted, could not support him. As a teenager he joined 43 Group, an organization set up by Jewish ex-servicemen who had fought in the Second World War and now clashed with antisemites and supporters of Oswald Mosley on the streets of London.[33] It was through 43 Group that Sassoon became aware of the option to join the Jewish struggle in the Middle East. A speech he heard by an officer who served in the Palmach, one of the Jewish paramilitary organizations that preceded the IDF, moved Sassoon and he decided to enlist. He traveled first to France, where he spent five weeks in a displaced persons' camp near Marseilles, awaiting transport to Israel. He later recalled hearing harrowing stories from some of the camps residents, many of whom were Holocaust survivors. "Having seen so much suffering at the camp, and having lived through so much racial hatred and bigotry as a child and young man, the intolerance showed by so many made me deeply angry," he wrote in his autobiography.[34] In contrast, once he arrived in Israel, Sassoon felt "that for the first time in my life I was somewhere where history was being made."[35] He joined an infantry unit and fought in the Negev Desert. Returning to Britain after the war, he went on to become a hairdressing tycoon.

Not all the foreign volunteers who fought for Israel's independence in 1948 felt as strongly about their Jewish identity as Sassoon. We have already seen how Harold Livingston from Massachusetts had only a vague perception of his Jewishness. Although technically a diaspora volunteer, he could also be categorized as a self-appointed ambassador because of his aforementioned desire to help the Jews of Palestine "fight for, and establish, a nation in the finest American tradition."[36] Nonetheless, participation in the war brought to the surface Livingston's identity dilemma as a diaspora member. Although he liked to tell himself that his "Jewishness [was] purely coincidental, an accident of birth," he later conceded that there was "a conflict of identity that had been raging within me for many years."[37]

The Armenian diaspora, much like the Jewish one, has a long history. Centuries-old communities increased in size and new ones were established when hundreds of thousands of Armenians were uprooted from their homes in Turkey during the First World War. The Bolshevik victory in the Russian Civil War led to the incorporation of parts of historical Armenia in the Soviet Union. With the decentralization of the Soviet Union in the late 1980s, pressure began to mount between Armenia and neighboring Azerbaijan. The region of Nagorno-Karabakh was a central bone of contention. Although it had been part of the Soviet Republic of Azerbaijan, its population was overwhelmingly Armenian. The situation deteriorated with the collapse of the Soviet Union. Armenia and Azerbaijan declared their independence from the Soviet Union in 1991, and soon a full-scale war erupted between them.

Armenia was able to benefit from the services of Armenian officers from various parts of the former Soviet Union, who had served in the Red Army, and now took on command roles in the Armenian military. Another source of support was the large Armenian diaspora, with members of geographically distant communities raising money and lobbying for political support in the United States and elsewhere. In addition to moral and financial support, approximately 1,000 diaspora members moved to Armenia, with a handful receiving high-level governmental appointments. Of the returnees, just under 200 volunteered for military service and fought in the war against Azerbaijan. These included Colonel Hovsep Hovsepian, who was born in France, moved to Armenia, and became a regiment commander; Monte Melkonian from the United States, who was killed in action in 1993; and Lieutenant Colonel Jirair Sefilian, a native of Lebanon who commanded irregular units in Nagorno-Karabakh.[38]

On the eve of the wars in the former Yugoslavia, there were a number of layers to the Croatian diaspora. Croats had settled in the United States, Canada, Australia, and other countries in a number of waves. First came the late nineteenth- and early twentieth-century economic migrants, then the political exiles who left Yugoslavia after the collapse of the pro-Axis Independent State of Croatia in 1945. In the late 1960s and early 1970s, after Yugoslavia liberalized its emigration policies, another wave of Croats moved abroad, with approximately 15,000 settling in Canada alone.[39]

Franjo Tudjman, the former communist partisan general who became a nationalist Croatian historian and a dissident, began to court diaspora Croat organizations in the late 1980s. These nationalistic organizations gave the aspiring leader political legitimacy, contributed funds to the party he established—the Croatian Democratic Union—and, in some cases, provided the cadre for Croatia's future leadership. According to the writer and political analyst Paul Hockenos, approximately 3,500 Croats from the diaspora returned to their historical homeland in the early 1990s to help build the nascent independent state (this number excludes the hundreds of thousands who resided within the borders of Yugoslavia and were displaced by the war). The most notable of these returnees was Gojko Šušak, who left Canada where he had been living for more than twenty years and arrived in Zagreb in 1990. Tudjman appointed him first as the minister of return and immigration, an indication of the new government's intention to foregather all exiled Croats. Then, from 1992 until his death in 1998, Šušak served as Croatia's defense minister.[40]

The Croatian diaspora also provided men who took an active part in the fighting. A number of these diaspora volunteers brought with them military experience from the French Foreign Legion or other national military forces. Ante Gotovina, for instance, had served as a paratrooper in the French Foreign Legion. Returning to his native Croatia when war broke out, he rose through the ranks to become a lieutenant general. Foreign volunteers without familial links to Croatia who served under Gotovina spoke highly of him.[41] Nonetheless, he was accused by the International Criminal Tribunal for the Former Yugoslavia of committing war crimes. In 2011 he was indicted for having taken part in the forcible removal of the Serb civilian population from Croatia's Krajina region between July and September 1995. However, this ruling was reversed in 2012, leading to Gotovina's acquittal.[42]

Another notable diaspora volunteer was Željko Glasnović, who, as a young man, had immigrated to Canada. Before fighting in the wars in the Balkans, Glasnović had served in the Canadian army. He later joined the French Foreign Legion and was posted to the Middle East during Operation Desert Storm in 1991. Leaving the legion before completing his service to fight for his native land, Glasnović became a colonel in the Croatian army. He was badly wounded in 1992 and dispatched a year later to Bosnia-Herzegovina to command the King Tomislav Brigade, a Bosnian-Croat unit. The American volunteer Rob Krott, who served under Glasnović in 1993, described his commander as inspiring "unswerving loyalty and intense confidence in his men."[43]

Kurdish foreign volunteers fall into two categories. Having no independent state of their own, most Kurds live in their historical homeland although the modern-day borders of Turkey, Syria, Iraq, and Iran divide them. Cross-border Kurdish volunteers from these four countries, who have participated in the PKK's struggle against Turkey or in the fight against the Islamic State in Syria and Iraq, are discussed below. For our present purpose, let us focus on the Kurdish diaspora. Financial constraints at home as well as repressive measures in Turkey have led a large number of Kurds to move abroad to countries like Germany. For some of these, their status as guest workers or refugees accentuated their identity as Kurds.[44]

For Milan Herekol (nom de guerre), the "diasporic turn" came in 1988, after the Iraqi dictator Saddam Hussein sent aircraft to bomb the Kurds of Halabja with chemical weapons. A member of a Kurdish-Australian family, Herekol grew up in Melbourne and worked for a local radio station. Whereas she did not know much about Kurdistan before the attack, news of what happened in Halabja changed her life, as she explained several years later: "From that moment I was only looking for a way to get to Kurdistan and help my people." In 1990 she traveled to Lebanon and met with the leadership of the PKK. She eventually decided to join the organization, arriving in Iraqi Kurdistan in January 1992. Remaining with the PKK for several years, in 2004 she told the war correspondent Peter Strandberg about the battles she fought in against the Turkish army and about the many friends she had lost on the way.[45]

As the examples above illustrate, a military crisis in the country of heritage could shape the actions of certain diaspora members, even in cases where the individuals in question did not have a particularly strong attachment to their co-ethnics beforehand. In cases where diaspora volunteers such as Glasnović spoke the local language and were familiar with local customs, they had a distinct advantage over foreigners who arrived in the warzone without such prior knowledge. Thus, they were more likely to serve in command positions. In some cases, when wartime military service ended, diaspora volunteers chose to settle down in their country of heritage. However, the postwar attitudes of members of this category were by no means homogenous. For some, like Livingston, their willingness to commit to the cause was tied to the specific circumstances created by the war. Once the war was over, Livingston, much like many other diaspora volunteers, returned to what he saw as his home country.

Cross-Border Volunteers

Cross-border volunteers, like those in the diaspora volunteer category, share ethnic or national ties with the group they enlist to fight for. However, while diaspora communities had been dispersed and live in two or more host countries, cross-border volunteers belong to populations that have remained in part of what they perceive as their historical homeland. Ethnic or national groups that have been divided by state borders are a persistent facet of modern history. Such divisions have often led to political tensions and international confrontations.[46] Cross-border volunteers put calls for political unity across borders into practice. They do not recognize, or at least seek to change, those state borders that cut through what is for them one ethnic or national group. Volunteers in this category often want to enlarge the borders of their home state, and in some cases also bring about internal change in its political structure, as the following examples illustrate.

Giuseppe Garibaldi's military volunteering throughout his long and illustrious career falls into a number of categories. His attempts to defend the Roman Republic in 1849 fit neatly into the cross-border volunteer category. After all, Garibaldi was not a Roman or a citizen of the Papal States, which the republic sought to replace. Technically speaking, he was a foreigner. After his return to Nice from South America in summer 1848 Garibaldi worked in tandem with Giuseppe Mazzini. Both men sought to do away with the borders that divided Italy and to unite the country under a republican regime. In July Garibaldi launched an appeal, calling on young people willing to fight for Italian unity: "Rush forward: gather round me: Italy needs ten, twenty thousand volunteers. . . . We will show Italy, Europe, that we want victory and we will have it."[47] He established a volunteer battalion under the slogan "complete independence or death" and, by the end of the month, marched with around 1,500 men to Bergamo, where he was joined by additional recruits. In the following weeks, Garibaldi's volunteers engaged Austrian troops around Lake Maggiore in a campaign that ultimately proved unsuccessful. Garibaldi was undeterred. He headed south and, in the spring of 1849, led the defense of the Roman Republic against troops dispatched from France. When addressing his troops, he gave voice to the belief that was self-evident to him but was still very distant in terms of the political reality: that Italy was one entity and that the Italians were one people. The gap between Garibaldi's beliefs and the political situation is perhaps best illustrated by the fact that, upon his return to his native Piedmont after the short-lived republic was crushed, he was arrested for entering the kingdom illegally. This was because he had allegedly forfeited his citizenship by fighting in the defense of Rome, a foreign state.[48]

As we have seen, ten years later Garibaldi chose Italian unification over republicanism and aligned himself with the Piedmontese leadership. But borders remained a sensitive issue for the leader of the Red Shirts. Garibaldi held a grudge against the Italian prime minister, Camillo Benso di Cavour, who had ceded his birthplace, Nice, to France in return for military support during the wars of Italian unification. In 1861 Garibaldi gave a speech at the Italian parliament, accusing Cavour of making him "a foreigner in Italy."[49] Furthermore, Garibaldi continued to struggle for the incorporation of Rome into the newly unified Italian state throughout the 1860s.

King Petar I and his prime minister, Nikola Pašić, the leaders of Serbia during the crisis of the summer of 1914 that culminated in the outbreak of the First World War, both shared a cross-border volunteering episode in their past. Nearly forty years earlier, in the summer of 1875, the future king, then known as Petar Karadjordjević, was in exile. While Serbia was ruled by the rival Obrenović dynasty, Petar trained in the French military academy at Saint Cyr and joined the French Foreign Legion to fight in the Franco-Prussian War of 1870. When the Serbian population of Herzegovina rebelled against the Ottomans in the summer of 1875, Petar went there to take part in the conflict. He led a band of about 200 men and fought under the alias Petar Mrkonjić.[50] The revolt in Herzegovina also animated Pašić, then still a relatively young Radical Party politician. Shortly after the outbreak of hostilities he felt that "the fate of all Serbdom depends in good measure on the outcome of the Hercegovina uprising."[51] In the autumn of that year he traveled to Herzegovina to take money that had been collected for their cause. He visited a rebel camp and met with local leaders, remaining in contact with them until 1876, when he returned to Serbia in time to take part in his home state's war against the Ottomans. In the 1890s Pašić continued to write at length about the need for Serbs and Croats to unite. His underlying assumption was that Serbs and Croats were essentially the same people.[52] The borders separating the south Slavs were artificial and therefore had to be removed. It was only in 1918 that this goal was achieved.

Like other Arab nationalists in the interwar period, Fawzi al-Qawuqji did not care much for the borders that were imposed on the Middle East at the end of the First World War. Born in Tripoli in northern Lebanon, he had served as an officer in the Ottoman army and fought against British forces north of Jerusalem in autumn 1917. This marked the starting point of his long career as a soldier of Arab nationalism. At various points in this career, which spanned more than thirty years, Qawuqji was a cross-border volunteer. When Syrian nationalists rose in revolt against the French Mandate in 1925, Qawuqji deserted the ranks of the French Troupes du Levant to play a leading role in the rebellion.[53] In 1928, after the revolt was crushed, Qawuqji's name was included in a blacklist, published by the French High Commission, of leaders who were prevented from returning to Syria.[54]

In the following years Qawuqji eked out a living training forces in Saudi Arabia and in the Iraqi military college. His next stint as a cross-border volunteer was in 1936, when the Arabs of Palestine rose in revolt against the British. Qawuqji crossed into Palestine, leading a band of about 300 foreign volunteers, and fought against British forces until he agreed to withdraw once the revolt was temporarily suspended in the autumn of that year. Qawuqji's disregard for the region's borders is exemplified by the statements he issued during his brief campaign in Palestine. He presented himself and his men as fighting in "Southern Syria" rather than in a separate and distinct country. Although the idea that Palestine was part of Syria soon disappeared from the Arab national discourse, in 1936 Qawuqji still firmly believed in it.[55]

In 1941 Qawuqji was in Iraq. He supported Rashid Ali al-Kaylani, the former prime minister of Iraq who led a coup against the pro-British government. Badly injured while fighting against British forces near Haditha in northwestern Iraq, he was airlifted to Germany where he spent the remainder of the Second World War. Qawuqji's final opportunity to act as a cross-border volunteer was in 1948, when he was appointed to lead the main cohort of the Arab Liberation Army, a volunteer force established by the Arab League to fight alongside the Palestinian Arabs in their war against the Jews.[56] As we shall see in Chapter 6, the Arab Liberation Army encountered several difficulties during the war of 1948, and the experience left Qawuqji despondent. For our present purpose, however, it is important to emphasize how this soldier of Arab nationalism viewed the struggles of the Arab peoples in the Middle East as part of one larger conflict. In an interview with the Lebanese daily *al-Hayat* in 1947, Qawuqji declared: "I have dedicated myself to the struggle to free the Arab lands from the yoke of foreigners, so I am a hostage to Arab events and interests. When the struggle calls me to whichever area I am needed, I am ready to answer that call."[57]

The borders separating the Arab states of the Middle East have also been challenged by a later generation of foreign volunteers. But while Qawuqji believed in a greater Arab nation that transcended the region's divisions, many of the cross-border volunteers of the 2000s and 2010s saw, and continue to see, their common ground in religious terms. There have been various attempts to ascertain the provenance of the foreign insurgents who joined militant Sunni organizations in Iraq post-2003. If the foreign suicide bombers that have been identified can be used as a barometer, then the largest cohort—more than 40 percent—was composed of Saudi nationals.[58] In the Iraqi context, as in other instances of cross-border volunteering, fighters from neighboring countries have a number of advantages over their counterparts from Western states. Aside from the geographical proximity, Iraqis and Saudis share the same language.

Doing away with the artificial borders allegedly created by the Anglo-French Sykes–Picot Agreement of 1916 is one of the key propaganda messages of some of the militant Sunni organizations in the Middle East.[59] This agreement was, and is still, held up as a symbol of the Western, colonial domination of the region—a domination that the organizations in question seek to reverse and destroy. Abu Musab al-Zarqawi, the Jordanian leader of Al-Qaeda in Iraq between 2004 and 2006, spoke in such terms in one of his propaganda videos: "We are not fighting for illusionary borders drawn by Sykes–Picot. Nor are we fighting to replace a Western tyrant with an Arab tyrant. Our jihad is more honorable than that. We fight to raise God's word on earth."[60] The same message also served the so-called Islamic State. The video *The End of Sykes–Picot* was released in late June 2014, after the organization captured Mosul and proclaimed the establishment of its caliphate. It showed bulldozers destroying the border rampart between Syria and Iraq in an area that was under the organization's control. Borders such as these, one Islamic State militant explained, fragmented the Islamic *umma* and were placed to "prevent the Muslims from travelling in their lands."[61] Quite aside from the symbolic and propagandistic value of this move, the Islamic State has been able to transfer Syrian fighters to Iraq and vice versa as well as to deploy its foreign fighters wherever they were needed on both sides of the old border. Of course, not all of the Islamic State's foreign fighters can be classed as cross-border volunteers. Some of them, as we shall soon see, fall within the substitute-conflict volunteer category. But Iraqi members of the organization, who were dispatched across the border by Abu Bakr al-Baghdadi in 2012 and 2013 to establish a stronghold in war-torn Syria, were cross-border volunteers par excellence.[62]

Ironically, the borders of the Middle East have been, and are still, under challenge from one of the main rivals of the Islamic State, the Kurds. Their ambition is to establish either independent or autonomous Kurdish entities in areas that currently belong to Turkey, Syria, Iraq, and Iran. This aspiration has led to cross-border deployments in practice, with Kurdish fighters being transferred between different fronts. For instance, in 2010 "Medya" (nom de guerre), a female PKK commander, was filmed operating with a small group of militants near the border between Turkey and Iraq. In 2014 the same commander was leading YPG troops fighting against the Islamic State in northeastern Syria, an area that the Kurds call Rojava. Some of the other Kurdish women and men fighting in Syria were born in Turkey.[63]

However, this type of cross-border mobilization does not appeal to all Kurds. The Kurds are divided not only by geography but also through language, clan and tribal affiliations, sectarian differences, and vying political standpoints. They have a long history of infighting. In fact, the authorities in Iraqi Kurdistan have, in a number of instances, prevented the passage of foreign volunteers from the territories they control into Rojava. This is because President Masoud Barzani and his Kurdistan Democratic Party, who have extensive links with Turkey, wish to placate Ankara, where concerns about the establishment of a new Kurdish entity in Syria run high.[64]

Another contested borderland that has been the site of intense fighting in recent years is the one between Russia and Ukraine. Like many other disputes, the present one could be seen to have various historical starting points. Nationalist Russian leaders have made claims for the Crimean Peninsula, Donbass, and southern Ukraine since the breakup of the Soviet Union in the early 1990s. In the case of Crimea, the Russian claim rests on the alleged illegitimacy of the transfer of the peninsula from the Russian Soviet Federative Socialist Republic to the Ukrainian Soviet Socialist Republic in 1954.[65] Promoted by Nikita Khrushchev, who was born in Ukraine and built much of his early career there, the "gift" of Crimea was supposed to mark 300 years of Russian and Ukrainian unification. Such territorial alterations between constituent republics were seen as a mere symbolic gesture at a time when the Soviet Union was at the height of its power. In any case, the leadership in Moscow authorized the transfer without consulting the local population. More broadly, a large part of the population in Crimea, Donbass, and other parts of eastern and southern Ukraine speaks Russian and has strong cultural ties with Russia. Their tendency to identify as Russian surged in 2014 as a result of the political turmoil in Ukraine. At the same time, the propensity to view parts of Ukraine as belonging to Russia also gained currency on the Russian side of the border.[66]

The identity and status of the armed Russians who fought alongside the separatists in eastern Ukraine in 2014–2015 was, and still is, a contested issue. For a long time Russia denied that its military was involved in the conflict, but the Ukrainians paraded captured Russian conscripts in front of reporters to prove otherwise. In December 2015 Putin conceded that Russian military intelligence officers operated in Ukraine but still denied the presence of regular troops.[67] Be that as it may, some of the pro-separatist Russian fighters in eastern Ukraine were not in active military service when the conflict erupted and could be classed as cross-border volunteers. One of these was Ruslan Starodubov. A veteran of Russia's wars in Chechnya, Starodubov held various jobs after being demobilized, working as a porter, a security guard, and a physical education teacher. When the conflict in Ukraine erupted in 2014, he enlisted to fight alongside the separatists. In early 2015 he was back in Russia, receiving treatment for a frontline injury he sustained. Describing himself and his fellow volunteers, he said, "We're normal nationalists. We love our people, and we love our homeland."[68]

Another Russian volunteer is a woman who identified herself as "Svetlana" when interviewed by reporters in early 2016. Before the war she worked in the publishing industry and had no previous military experience. In the summer and early autumn of 2014 she paid close attention to news coverage of the conflict in Ukraine and sent humanitarian aid packages to assist the separatists' cause. "All that was happening touched me—resurgent fascism, the genocide of the Russian population, attacks on civilian towns—and I could not remain indifferent," she explained. Finally, Svetlana decided to leave her job and to volunteer for military service, arriving in Donetsk on 1 December 2014.[69] By volunteering to fight in Ukraine, both Svetlana and Starodubov saw themselves as resisting Russia's enemies and assisting fellow nationals who happened to reside beyond their state's borders.

Different cross-border volunteers have positioned themselves in different ways in relation to the governments of their home state. Qawuqji in 1925 and 1936 and al-Zarqawi in 2004–2006 sought to topple the regimes that controlled the Middle East. Conversely, Starodubov's actions were aligned with the policy of his government. The one thing that the individuals discussed here had in common, despite the vast differences in their historical circumstances, was their willingness to fight alongside co-ethnics or co-nationals across the border and, ultimately, to alter the map of their respective regions.

Substitute-Conflict Volunteers

Some of those who come under the substitute-conflict volunteer category see themselves as patriots of their home state, much like the self-appointed ambassadors. However, substitute-conflict volunteers are sworn enemies of their own governments. They consist of dissidents and political émigrés who seek to change the regimes in their home states but have not been able to attain their goal. Instead, they enlist to fight in a war elsewhere in the hope that one victory will pave the way for the one they long for. We have already seen how Jan Henryk Dabrowski, one of the founders of the Polish Legions that fought for Napoleon in Italy, tried to convince the French leader to allow the Poles to march on Poland and fight for its liberation. The Polish exiles hoped that by joining forces with France, the archrival of the powers that partitioned Poland, they would gain an opportunity to win back the independence of their homeland.[70] Their hopes were dashed, as were those of other substitute-conflict volunteers who pursued a similar path in subsequent generations.

The largest and best-organized contingent of foreign volunteers who fought for the restoration of Bourbon rule in Sicily and southern Italy in the early 1860s were Carlist exiles from Spain. The Carlists sought to establish a separate line of the Bourbon dynasty, that of the descendants of Don Carlos, on the Spanish throne. To achieve their goal, they had fought against the forces of Queen Isabella II throughout most of the 1830s and again in the late 1840s. In terms of European politics, the Carlists stood for traditional values and legitimist rule. However, by 1860 their prospects of success in Spain had reached a nadir. Émigré Carlists who were based in France such as generals José Borges and Rafael Tristany sided with Francis II of Naples and fought against the unification of Italy. Carlist leaders, who were partly motivated by the desire to escape their financial difficulties, spoke in terms of counterrevolutionary solidarity. They saw the struggle in southern Italy as a continuation of their fight in Spain. Francesc Tristany wrote to his brother, Rafael, in February 1861:

> We have no aspirations other than to continue to make those sacri-
> fices, to the day we give our lives, if necessary, to defend the cause
> of our King. The sword that we brandished in Spain, we shall draw
> again to fight in favor of legitimacy wherever it becomes necessary:
> the revolutionaries are the same everywhere, and their plans are al-
> ways iniquitous. The usurpation that has been committed to the
> detriment of the King of Naples cries out for deserved vengeance,
> and we consider it a great honor to lend our hand.[71]

After landing in Calabria, the volunteers from Spain fought a guerrilla campaign against forces loyal to the government of newly unified Italy. Borges was captured and executed in December 1861. Meanwhile, Tristany launched small-scale attacks in the Abruzzi. By 1863 he was forced to depart, taking with him most of the Carlists who were still present in Italy.

Substitute-conflict volunteers continued to move within the Italian-French-Spanish triangle in the twentieth century, though in the opposite direction. The outbreak of the Spanish Civil War in the summer of 1936 spurred the Italian antifascists who resided in France into action. These émigrés were not a homogenous group. Some were socialists, others were anarchists, and others still were communists. Carlo Rosselli, the leader of the antifascist resistance movement Giustizia e Libertà, began to recruit Italian émigrés of different political hues to fight in the civil war in Spain in late July. By early August the first volunteers began making their way toward Barcelona, where they were soon joined by Rosselli. Much like the German exile Gustav Regler, Rosselli later recalled that, "after the long years of exile, I must confess that it was only when I crossed the border of Spain, once I enrolled in the popular militias . . . and embraced the rifle, I felt that I became a free man."[72] While the struggle in Spain gave Rosselli and other émigrés a sense of purpose, they saw their participation as a means to an end, a possible vector toward their final goal. As Rosselli declared in his famous appeal on Radio Barcelona on 13 November 1936, the Italian volunteers were fighting "today in Spain, tomorrow in Italy."[73]

The Germans who joined the International Brigades in Spain also belong to the substitute-conflict volunteer category as far as their relationship with their home state is concerned. Like their Italian counterparts, they were drawn essentially from among political exiles who had left Nazi Germany. Historian Josie McLellan has aptly described their struggle as "a displaced fight against Hitler and a chance to strike a blow against international fascism."[74] The lyrics of Ernst Busch's *Ballade of the XI Brigade* echoed this sentiment:

> And even if we have to fight
> For seven more years,
> Every war's over sometime.
> We're going to see Germany again![75]

To an extent, both the Italian and the German volunteers in the International Brigades were not only fighting Franco, an ally of Mussolini and Hitler, but the actual forces of their home states as well. The German Condor Legion, dispatched to Spain by Hitler, provided between 14,000 and 19,000 military personnel to support the Spanish Nationalists. The Corpo Truppe Volontarie and other Italian forces committed to Spain by Mussolini numbered more than 70,000 troops.[76] In the Battle of Guadalajara in March 1937, Italian troops sent by Mussolini and antifascist Italian volunteers in the International Brigades physically fought each other (with the antifascists prevailing). As one of the commanders of the International Brigades pointed out in a report to his superiors in Moscow, the German and Italian volunteers were "fighting with arms in hand, that is, in the most effective and revolutionary way, first and foremost against their own enemy, which has already oppressed their own countries and peoples for many years."[77]

While the Condor Legion and the Corpo Truppe Volontarie, both sponsored by their governments, are beyond the scope of this book, there were nonetheless substitute-conflict volunteers fighting on Franco's side. Historian

Judith Keene estimates that between 80 and 100 White Russian émigrés served in the Spanish Foreign Legion, in Carlist regiments, and various other Nationalist units during the civil war. The Russian Armed Services Union, a White Russian veterans' organization in France, embraced Franco's claim that his cause was a "crusade against communism." Émigré officers, who had been defeated in the Russian Civil War less than two decades previously, hoped to use the war in Spain as an opportunity to regroup, deliver a blow against Bolshevism, and create a new imperial army in preparation for when they could return to Russia.[78] Although the Russian volunteers who fought for the Nationalists were not always received by their Spanish hosts with the respect they expected, they continued to emphasize how the two causes were aligned. In July 1937, for instance, General Anton Nikolai Shinkarenko sent a long petition to Franco in which he pointed out that White Russia and Nationalist Spain shared "the noble cause of defending European Christian culture." He also stressed that he and his fellow émigré officers had been fighting "our mortal enemies," the Reds, since 1917.[79] Like other volunteers belonging to the substitute-conflict category, the White Russians fitted the foreign cause they wished to fight for into the political frameworks of their home state.

The Cold War, which saw the struggle between Left and Right globalized on an unprecedented scale, also produced substitute-conflict volunteers. When Che Guevara crossed Lake Tanganyika into eastern Congo in April 1965, he discovered that some of the troops engaged in the revolutionary struggle he had come to assist were Rwandan. He soon noted that relations between the Congolese and the Rwandans were strained. The commander of the Rwandan troops, Mundandi, often complained to Guevara about the "lack of fighting spirit" among the Congolese, and about how he would be left "without men to make the revolution in Rwanda."[80] In November 1965, as the situation was becoming desperate and the local rebels along with their Cuban and Rwandan supporters were in retreat, Mundandi wrote to Guevara saying, "I have tried to help this revolution so that it will be possible to make another one in our country. If the Congolese will not fight, I prefer to die on soil intended for the Rwandan people. If we die on the way, that is alright too."[81]

Some of the foreign fighters in the civil war in Syria and in the conflict in eastern Ukraine can also be categorized as substitute-conflict volunteers.

This is certainly the case with the anti-Russian Chechen volunteers who have fought in both conflicts. The First and Second Chechen Wars of the 1990s paved the way for Ramzan Kadyrov, an ally of Russia's Vladimir Putin, to install himself as head of the Chechen Republic. Separatist and jihadist Chechens, who opposed the alignment with Russia, went into hiding or fled abroad. In November 2015 Abdulvakhid Edelgireyev, a key figure in the anti-Russian Chechen organization the Caucasus Emirate, was shot dead in Istanbul. His assassination brought to an end the struggle of a highly committed substitute-conflict volunteer. In 2013 Edelgireyev and a group of Chechens living in Istanbul left to join the civil war in Syria. He pledged allegiance to Jabhat al-Nusra, an affiliate of Al-Qaeda, and remained in Syria for several months.[82] Other Chechen volunteers joined the Islamic State. In September 2014 the organization released a video showing their fighters in a Syrian air force base they had captured shortly beforehand. Some of the fighters were Chechen, and the video showed them issuing the following threat: "Vladimir Putin, these are the Russian planes that you sent to Bashar [al-Assad]. Allah willing, we will take them and liberate Chechnya and the Caucasus."[83]

But Edelgireyev became disillusioned with the chaos in Syria. As his father later explained, "our struggle was always about Russia, he wasn't interested in Isis." Hence, in 2015, Edelgireyev traveled to Ukraine to fight alongside forces loyal to the government in Kiev against the Russian-backed rebels. After spending three months in their midst, the Ukrainians asked Edelgireyev to leave, a move that Edelgireyev's family understood as designed to placate the West.[84] Nonetheless, there were other Chechen volunteers fighting in eastern Ukraine. One of them, a commander of a volunteer Islamic battalion in Mariupol who wished to remain anonymous, told the New York Times in 2015: "We like to fight the Russians. . . . We always fight the Russians." He said he had been on the same path for 24 years, since the breakup of the Soviet Union. As a veteran of the Chechen wars of the 1990s, the commander declared: "The war for us never ended. We never ran from our war with Russia, and we never will."[85] Similarly to their predecessors in the civil war in Spain, the Chechen substitute-conflict volunteers who fought on the side of the Ukrainian government faced fellow Chechens on the opposite side. In December 2014 Reuters reported on a 300-strong unit of Chechens loyal to Russia, fighting alongside the east Ukraine separatists.[86] Of all the categories, it seems that substitute-conflict volunteers are the most prone to encounter co-nationals among their enemies on the battlefield.

Substitute-conflict volunteers, whether they were Poles fighting for Napoleon or Chechens fighting in Syria and Ukraine, convinced themselves to overlook any differences that might have existed between their own cause and the one they joined. For instance, the White Russians, who were devout Orthodox Christians, were willing to gloss over the adherence of the Nationalists in Spain to Catholicism for the sake of fighting their common enemy, communism. Muslim Chechen volunteers in Ukraine fought on the same side as Far Right Western European volunteers just so they could strike a blow against Putin's Russia.

* * *

By joining a foreign conflict without being sent by their government, the volunteers also express their opinion on the foreign policy of their home state. In general, self-appointed ambassadors seek to bring their governments around to supporting one of the sides in a foreign war. When diaspora volunteers commit themselves to military service abroad, they declare, either expressly or inadvertently, that in that specific point in time their affinity to their co-ethnics overrides their commitment to their state of residence. Cross-border volunteers want to enlarge their home state so that it will include co-ethnics or co-nationals who are stranded, so to speak, on the other side of an artificial state boundary. This desire for territorial expansion coincided, in some cases, with an aspiration to change the government of their home state. And, finally, substitute-conflict volunteers openly seek to overthrow the regime of their home state and view participation in a foreign conflict as a means of achieving this goal.

The same conflict could attract foreign volunteers who belong to different categories. For instance, self-appointed ambassadors and substitute-conflict volunteers fought side by side in the International Brigades during the Spanish Civil War. In Syria, both cross-border and substitute-conflict volunteers joined the Islamic State. In other words, volunteers of varying national backgrounds could enlist to join the same cause even when their stance vis-à-vis their respective home states differed substantially.

From the perspective of the home state, volunteers belonging to the substitute-conflict category pose the greatest threat. For instance, Italian and

German antifascists fighting in Spain also sought to topple the regimes of Mussolini and Hitler back at home. Cross-border volunteers were only dangerous for the governments of their home states in cases where they hoped to bring about regime change to coincide with territorial expansion, as Garibaldi did in 1849. Self-appointed ambassadors and diaspora volunteers may pose a challenge to their home states' neutrality and therefore have an effect on the foreign policy of those states. But their ambition for change revolves around the fate of the host state rather than a regime change at home. As we shall see in the next chapter, governments have sometimes had an intuitive understanding of the subtle differences between different types of foreign volunteers. However, in most cases, the policies of home states have been driven by other interests and considerations.

* 5 *

Controlling the Flow

Governmental Responses, Legislation, and Support Networks

Challenging the State

Foreign volunteers pose a challenge to their home states and to the international system. Because the causes they join are often contentious, the departure of volunteers could be politically divisive and perhaps even create domestic difficulties for their governments. At the same time, volunteers may negatively impact the foreign affairs of their home state by compromising its neutrality and breaching its international obligations.

As we have already seen, the expectation that citizens should serve only in the armed forces of their own country developed gradually from the late eighteenth century onward. Such an expectation hinges on the principle of loyalty. The scholar of nationalism Anthony D. Smith attributed this loyalty to what he called the "modernist paradigm of nationalism": "Nations constituted the primary political bond and the chief loyalty of their members. Other ties of gender, region, family, class and religion—had to be subordinated to the overriding allegiance of the citizen to his or her nation-state."[1] According to this approach, only national loyalty could mobilize the masses "for the commitment, dedication and self-sacrifice required by modernisation with all its strains and dislocations."[2] Modern governments, therefore, resented the idea that their sovereignty and monopoly over military recruitment might be infringed by citizens acting in the interest of another state or political entity.

The French Revolutionary Wars of the 1790s provide an early example of such tendencies. In 1793 Edmond-Charles Genêt, the new French minister to the United States, tried to recruit American citizens to man French vessels and to fight against France's enemies. President Washington disapproved, issuing in April of that year a proclamation titled "Neutrality of the United States in the War Involving Austria, Prussia, Sardinia, Great Britain and the United Netherlands against France." The proclamation forbade citizens from aiding or abetting hostilities against any of the powers involved in the conflict.[3] In June Thomas Jefferson, the secretary of state, explained to the French minister that "granting military commissions, within the United States by any other authority than their own, is an infringement on their sovereignty, and particularly so, when granted to their own citizens, to lead them to commit acts contrary to the duties they owe their own country."[4] This policy was further cemented by the Neutrality Act of 1794, which outlawed the raising of an unsanctioned military force within the jurisdiction of the United States. It also criminalized engaging such a force against a state at peace with the United States. Underlying this policy was the fear that the actions of individuals might drag the still weak United States into the conflict that had erupted in Europe. Possible domestic strife between partisans and opponents of the French Revolution was also a concern.[5]

The political scientist Janice E. Thomson described the act of 1794 as a watershed in the development of the concept of neutrality, a policy that was emulated by many other states in the decades that followed. Between 1794 and 1938, forty-nine states enacted some form of legal control over their citizens' ability to enlist abroad. Many more countries have done so since. States therefore allowed themselves to exercise greater control over the activities of their citizens. Thomson argued that, as this norm in international affairs became more prevalent, states found it increasingly difficult to recruit military personnel abroad. Writing in 1990, before the giant rise in the use of private military and security companies, she pointed out that "nearly 90 percent of the world's armies recruit exclusively within their home states' territories."[6]

There is evidence, primarily from the nineteenth century, to support Thomson's argument about states adopting legislation to curtail foreign enlistment as a result of pressure from belligerents in a foreign conflict. In the late 1810s, when the Spanish American territories fought for their independence from Spain, the latter put pressure on both the United States and Britain to observe their neutrality and prevent their citizens and subjects from joining the insurgents. In March 1817 and again in April 1818, the US Congress passed legislation that extended the Act of 1794 to wars that involved a "colony, district or people" and not only recognized independent states.[7] In Britain the Foreign Enlistment Act was passed in 1819. It prohibited British subjects from enlisting without authorization in the military services of any "foreign prince, state, potentate, colony, province or part of any province or people, or of any person or persons exercising or assuming to exercise the powers of government in or over any foreign country."[8]

In the 1860s the Civil War in the United States created new international disputes about neutrality and the involvement of individuals from non-belligerent states. The CSS *Alabama* was built near Liverpool in Britain for the navy of the Confederacy in 1862. It raided several Union merchant vessels across the globe before being sunk in 1864. The government of the North accused Britain of breaching its neutrality by allowing Confederate ships to leave British shipyards. To make matters worse, the majority of the seamen on board the *Alabama* in 1863 were Englishmen. Legal disputes between Britain and the United States regarding the "Alabama Claims" continued for a number of years after the Civil War ended. Eventually, following international arbitration, the British had to pay $15 million in compensation. Largely as a result of this and other incidents during the Civil War, the British government passed a more stringent Foreign Enlistment Act in 1870, to protect itself from similar claims in the war that broke out between Prussia and France.[9]

But pressure applied by belligerent states was not the only factor that shaped the ways governments responded to foreign volunteering. Thomson's emphasis on developments in international law and the obligations of neutral states overlooks the ways in which foreign war volunteers have affected the domestic politics of their countries. From the point of view of governments seeking to remain on the sidelines of conflicts abroad, such volunteers could pose an altogether different risk. If foreign volunteers attach themselves to a cause that enjoys a degree of popularity, their efforts and sacrifices could raise support for one of the sides in the conflict to such an extent as to force the government's hand to intervene.

According to historian David MacKenzie, "from July to November 1876, General Mikhail Grigorevich Cherniaev was unquestionably the most popular man in Russia."[10] As we have already seen, he sought to champion the Pan-Slavic cause and volunteered his services to help the Serbs fight against the Ottoman Empire. Meanwhile, the government of Tsar Alexander II preferred not to enter the conflict in the Balkans in the autumn of 1876. A war with Turkey, the finance minister warned, could mean financial ruin for Russia. There was also uncertainty about the response of the other great powers in Europe. At the same time, the government faced enormous public pressure to assist Serbia. Russian statesmen, even those who were not Pan-Slavs, increasingly felt that Russia needed to uphold its national honor by intervening on behalf of the Orthodox Christians of the Balkans.[11] Although militarily the efforts of Cherniaev were a failure, he and the other Russian volunteers helped to direct public attention to the crisis. By late 1876 the tsar himself spoke of "our volunteers who have paid with blood for the cause of Slavdom."[12] In February 1877 Alexander II declared: "In the life of states just as in that of private individuals there are moments when one must forget all but the defense of his honor."[13] Two months later Russia declared war on Turkey.

Another instance where foreign volunteers fanned support for intervention in a conflict abroad was in the United States during the First World War. American authors and intellectuals who were involved in voluntary medical activity in France and Belgium aroused emotional sympathy for the Allies.[14] Meanwhile the volunteer pilots from the United States who fought for France stirred up support for intervention. The thirty-eight American airmen of the Lafayette Escadrille and other volunteers from the United States in the French armed forces during the Great War could be categorized as self-appointed ambassadors. One of the veterans of the squadron, James Norman Hall, later claimed that "the outstanding accomplishment of the volunteers was their influence on public opinion in America at a time when we were neutral."[15] Writing from a less enthusiastic point of view in 1940, the attorney and sociologist David Riesman Jr. warned of the danger that "our country would go to war again in the wake of an advance guard of adventurous or idealistic volunteers in a belligerent force."[16]

Hence, foreign policy considerations and international obligations do not account, on their own, for the policies of home states toward foreign volunteering. Indeed, in several cases it was primarily domestic politics and considerations of political expediency that governed whether and to what extent states acted to preclude the foreign enlistment of their citizens.

Political Expediency

While several countries have adopted legislation limiting or prohibiting foreign enlistment to keep in line with international norms of neutrality, the application of such laws has always been subject to domestic considerations of political expediency. As the following survey of historical cases from nine different countries illustrates, the international norm that Thomson described was and still is far less prevalent than legal theory might suggest. First of all, the strictness of the legislation varies from state to state. Moreover, when it became politically expedient, some home states allowed and even facilitated transnational volunteering.

Legislation curtailing foreign enlistment was, and is still, normally passed in response to a specific need. The criminal code of Bosnia-Herzegovina was only amended in June 2014 following growing concern about the participation of Bosnian foreign fighters in the civil war in Syria. The new law criminalizes service in or recruitment for "foreign military, foreign paramilitary or foreign parapolice formations." Enlisting, directing, or equipping citizens of Bosnia-Herzegovina is punishable by imprisonment for a term of no less than five years, while actually joining a foreign force carries a prison sentence of no less than three years.[17]

In contrast, Israel has no ban on foreign enlistment as such, presumably because the political pressure to pass such legislation has never been sufficiently strong. Prosecutions against foreign volunteers are based on terrorism or security-related charges. The example of 23-year-old Ahmed Shurbaji is typical. Coming from an Arab town in northern Israel, Shurbaji and two friends traveled to Turkey in January 2014 and then were smuggled into Syria. He initially joined an organization called Jaysh Muhammad but soon transferred to the Islamic State. During his time in Syria he underwent firearm and basic military training, attended religious classes, was posted on guard duty, and participated in two battles. In April of that year he decided to leave and notified Israeli authorities of his intention to return. He was arrested upon arrival. Shurbaji was charged with illegally traveling to Syria and undergoing forbidden military training. The latter charge is part of the 1977 penal code that deals with sedition. He was not charged with joining a terrorist organization because Israel designated the Islamic State as an unlawful association only in September 2014, after Shurbaji had already returned. The judge convicted Shurbaji, accepting the prosecution's argument that the military knowledge he acquired could be used to carry out terrorist activities in Israel. He received a twenty-two-month prison sentence, which was justified as necessary to deter others from pursuing a similar path.[18]

As in so many other countries, the decision whether to prosecute seems to depend more on the identity of the foreign volunteers and the cause that they joined than on the fact that laws had been broken. The Canadian-Israeli Gill Rosenberg participated in the same conflict as Shurbaji, but she served with Kurdish forces fighting against the Islamic State in Syria and Iraq. After returning to Israel in 2015, the General Security Service questioned her, but she was not detained despite having broken Israeli law by entering an enemy state.[19] This selective approach toward prosecution was and still is practiced by countries with long traditions of legislation to counter foreign enlistment.

Sweden's twentieth-century record suggests that its approach to foreign enlistment was conflict-specific. Eager to preserve its neutrality in an increasingly polarized Europe, the Swedish government passed the Act on Measures to Prevent Volunteers from Participating in the Spanish Civil War in early 1937. Nevertheless, approximately 550 Swedes joined the International Brigades while fewer than ten fought on the Nationalist side. None of the volunteers who returned from Spain were prosecuted.[20] In contrast, no similar legislation was passed during the Russo-Finnish Winter War. As we have seen, the Swedish government faced a serious predicament in late 1939, after the Soviet Union invaded neighboring Finland. Public support for the Finnish cause soared, but the government had to tread carefully. It could not afford overly to antagonize the Soviet Union but also had to be mindful of Germany, which was engaged in a war with Britain and France that could extend to Scandinavia. Moreover, Berlin had recently signed a non-aggression pact with Moscow, which made it more difficult to guess Hitler's future intention. As the British journalist Geoffrey Cox observed, Stockholm "preferred to balance on the slippery pole of neutrality."[21] Hence, in most of its assistance to Finland, the Swedish government stopped short of direct military intervention.[22]

In order not to jeopardize Sweden's defenses, the government capped the number of volunteers that would be given permission to leave at 8,000, raising the limit to 12,000 in February 1940. Pilots who wished to serve in Finland asked to be released from the Swedish air force on the understanding that they would be re-hired upon their return. Cox reported that "In some barracks commanders called their men on to parade and gave them a chance to volunteer."[23] However, men from the army who joined the Swedish Volunteer Corps could not wear their regimental uniform. Instead they received a new type of uniform with insignia that was created especially for the conflict in Finland. In January 1940 the Soviet envoy in Stockholm protested against the recruitment of volunteers for Finland. The Swedish government replied that it had no responsibility for the actions of private citizens who went to Finland on an entirely voluntary basis.[24]

Sweden's approach toward the recruitment of Swedes to the German Waffen SS during the Second World War was much more hesitant. Following the German invasion of the Soviet Union in summer 1941, Stockholm offered Berlin to send Swedish officers to the eastern front. However, once this offer had been turned down, the Swedish government decided in early September 1941 to forbid any of its citizens from serving in any foreign army except the Finnish.[25] Some 200 to 300 Swedes joined the Waffen SS nonetheless. After the war was over, Sweden—in the words of Martin Gutmann—"chose to forgive and forget." Most of the returning volunteers were ignored. Only those who had deserted from Swedish military units, stolen military property—such as their Swedish army uniforms—or leaked sensitive information to German intelligence were prosecuted.[26]

Canada too has a record of selectively implementing its foreign enlistment legislation. During the American Civil War, Canadians—or, more precisely, British North Americans—were still bound by the British Foreign Enlistment Act of 1819. Nonetheless, several thousand British North Americans served in the armed forces of the North while a far smaller number fought for the South. The exact number of volunteers cannot be determined because of the diverse methods in which they enlisted and because the US War Department's record keeping regarding the nativity of soldiers was very poor. There were, however, some exceptional cases. In 1861 Colonel Arthur Rankin, a member of the Canadian Parliament and the commander of the Ninth Military District of Canada, offered his services and was subsequently commissioned by the US government to raise a regiment of 1,600 lancers for the army of the North. When he requested leave of absence and began to set his plan in motion, Rankin was arrested for violating the Foreign Enlistment Act. He was never convicted, though he was forced to relinquish his American commission and was deprived of his Canadian commission as well.[27]

Canada passed its own Foreign Enlistment Act in 1937. In doing so, the government of Mackenzie King was not only complying with the international Non-Intervention Agreement concerning the war in Spain (to which we shall return later on) but also trying to offset potential domestic political tensions. While English-speaking Canadians were, broadly speaking, sympathetic toward the Spanish Republic, public opinion in Quebec tended to be much more sympathetic toward Franco, especially in circles close to the Catholic Church. The Canadian Foreign Enlistment Act was therefore designed to placate Quebec voters without alienating too many English-speaking Canadians.[28] Political considerations aside, the Foreign Enlistment Act did not give Canadian authorities an effective tool to prevent the departure of volunteers to Spain. In September 1937 the commissioner of the Royal Canadian Mounted Police, James Howden MacBrien, asked the ministry of justice for direction:

> It would be appreciated if we could be advised as to whether you desire that the provisions of the Foreign Enlistment Act should be strictly enforced. . . . The whole question regarding the action to be taken in connection with the Foreign Enlistment Act appears to be entirely dependent upon the wishes of the Government.[29]

As the government was still hesitant, MacBrien's assistant informed his subordinates that it was "considered desirable at present to allow such persons as those mentioned to proceed to their destination, until an opportunity arises whereby a prosecution or prosecutions can be instituted."[30]

The Canadian government's adherence to a policy of preventing foreign enlistment has been limited also in subsequent conflicts. The Canadian Ministry of External Affairs determined in 1947 and 1948 that the Foreign Enlistment Act was not applicable to the conflict in Palestine. However, Canadian Jews who volunteered to assist the nascent Jewish state operated under the assumption that volunteering to fight for Israel was illegal and thus carried out their activities with the greatest secrecy.[31] In 2014, once the question of state policy vis-à-vis Canadian citizens fighting alongside Kurdish forces in Syria and Iraq was raised, the government has been accused of sending mixed signals. Public Safety Minister Steven Blaney told CBC News that the government "would not oppose a citizen who is willing to engage in a battle for liberty and helping the victims of barbaric crimes." At the same time, the chief of defense staff, Tom Lawson, said he did not "encourage Canadians to leave our nation and head to other nations to get involved with the militaries of that nation."[32]

Historically, Britain's policy regarding the prevention of foreign enlistment has been very flexible. To begin with, the British Foreign Enlistment Act does not prohibit all forms of foreign enlistment, limiting itself to service in wartime against a government at peace with His or Her Majesty. When wars were fought between independent states—such as the Russo-Turkish War of 1877–78, the Sino-Japanese War of 1894–1895, and the Russo-Japanese War of 1904–1905—the Foreign Enlistment Act was applied, with British authorities preventing the sailing of naval vessels to belligerents in these conflicts. Civil wars and wars of secession, however, elicited ad hoc and inconsistent responses from the government. Part of this inconsistency derived from the British government's uncertainty about whether or not the insurgents in a given conflict should be afforded recognition. Often dilemmas also arose from the state of play in international relations. For instance, in an attempt to show other powers in Europe that it was adhering to its policy of neutrality during the conflict in Greece, London issued proclamations in 1823, 1824, and 1825, calling attention to the provisions of the Foreign Enlistment Act.[33] A number of British volunteers in Greece who had been officers before the conflict had their names struck from the army and navy lists.[34] However, there were no prosecutions.

The British government's response to the First Carlist War in Spain (1833–1839) was completely different. This civil war revolved around the question of who would succeed to the throne of Spain. It also pitted liberal reformers, who supported the wife of the late Ferdinand VII, Queen Cristina, and their daughter, Isabella, against ultra-absolutists who congregated around Ferdinand's brother, Don Carlos. Britain backed the queen and the liberals diplomatically and through the supply of arms. However, in 1835, the Carlists enjoyed military success and the Spanish ambassador in London asked the British government to send military assistance in the form of a force of 10,000 volunteers. The British government agreed. To facilitate the recruitment of such a force, a King's Order in Council on 10 June 1835 temporarily suspended the Foreign Enlistment Act for a two-year period specifically for those who wished to "engage in the military and naval service of Her Majesty Isabella the Second, Queen of Spain." This move paved the way for the establishment of the British Auxiliary Force, which saw action in Spain later that year.[35]

British volunteers were again drawn to a civil war in Spain just over a hundred years later. By this stage the technology of warfare and the scale of the conflict had changed, as had the international context and the ideologies under dispute. However, the government's inability to formulate a consistent policy with respect to foreign enlistment in civil wars remained a feature of the twentieth century. Officials in the Foreign Office sought to limit the flow of volunteers from Britain to Spain. Some of them had ideological misgivings about the Spanish Republic but, more importantly, they wanted to be in a position to pressure other European states to stop their military involvement in the conflict. Evelyn Shuckburgh, who was responsible for Spanish affairs at the Foreign Office, worried that Britons fighting in Spain were "certain ultimately to be a responsibility to us, and they may involve us in complications."[36] On 11 January 1937 the government declared through a press notice that, in accordance with the Foreign Enlistment Act, it was illegal to recruit or volunteer for the armed forces of all sides in the Spanish conflict. Those convicted would be liable to a prison sentence of up to two years or a fine or both. Subsequently, a few enlistment-related cases were referred to the Director of Public Prosecutions, but as on previous occasions, no legal action was taken.[37] The Home Office was pessimistic about the chances of successful prosecution. Furthermore, the government's declaration of January 1937 did not stop the flow of volunteers to Spain. Although numbers dropped from around 200 in December 1936 to about a dozen in March 1937, the pace later intensified. Roughly two-thirds of those who left Britain to fight for the Spanish Republic did so after the declaration.[38]

The Soviet Union's attack on Finland in late November 1939 was met with a very different response by the British government. Many high-ranking British officials, led by Sir Samuel Hoare, the Conservative Lord Privy Seal in the War Cabinet, were sympathetic toward the plight of the Finns. Ironically, Hoare found inspiration in the armed support that Nazi Germany and Fascist Italy had provided to Franco three years earlier during the Spanish

Civil War. On 4 January 1940 Hoare suggested to the cabinet that "we should examine the possibility of giving assistance [to the Finns] on the Spanish precedent, but with the difference that personnel sent to Finland should be true volunteers and not recruited from the serving ranks of the Regular Army."[39]

The British government did not obstruct, and in some cases assisted, a recruitment office for volunteers for Finland that was set up by Colonel Harold Gibson, an employee of the Cabinet Secretariat. Leading political figures such as Leo Amery, the future secretary of state for India, and Winston Churchill, then the First Lord of the Admiralty, were either members or supporters of the Finnish Aid Bureau.[40] A question was subsequently raised at the House of Commons in February 1940 about whether the government intended to apply the Foreign Enlistment Act as it had done during the Spanish Civil War. The government's written response was that Britain was merely complying with a League of Nations resolution, which had been adopted in December 1939. Even though it had become completely ineffectual by this stage, the League called on member states to "provide Finland with such material and humanitarian assistance as may be in its power, and to refrain from any such action which might weaken Finland's powers of resistance." Hence, the British government stated that "it would, in their view, be inconsistent with the spirit and with the terms of that resolution that British subjects who wish to volunteer for service in Finland should be hindered by the provisions of the Foreign Enlistment Act."[41] The difference in the government's policies during the Spanish Civil War in 1937 and Soviet-Finnish War in 1940 highlights the extent to which the implementation of the act was subject to political and ideological considerations.

In 1976 Prime Minister Harold Wilson considered amending the Foreign Enlistment Act following the participation of British mercenaries in the civil war in Angola. While eventually the act remained unchanged, Wilson's comments at the House of Commons illustrated the extent to which the government's approach toward foreign enlistment was cause-dependent. He pointed out that "there are different occasions, reasons, motives and inspirations for people going abroad to fight." For him, the desire of British Jews to travel to Israel, "the land which is the foundation of their faith," in 1973 to participate in the Yom Kippur War against Egypt and Syria was "understood by everyone."[42]

In the early 2000s Britain adopted a number of antiterrorism measures in the wake of the global "war on terror." The Terrorism Act of 2006, for instance, allowed the government to prosecute those who had received training in terrorist techniques or attended terrorist training camps. These new measures made the Foreign Enlistment Act redundant. When British foreign fighters returning from the civil war in Syria were prosecuted, they were charged under the various Terrorism Acts rather than for breaching the Foreign Enlistment Act.[43] Britain's selective approach to the prosecution of former foreign volunteers is succinctly summarized in a statement to the press issued by the Home Office in 2014: "UK law makes provisions to deal with different conflicts in different ways—fighting in a foreign war is not automatically an offence but will depend on the nature of the conflict and the individual's own activities."[44]

Having secured its independence from Britain, neighboring Ireland (Éire) has had a similarly flexible approach toward foreign enlistment. Here too internal politics and foreign policy considerations influenced the government's policy. In early 1937 the Irish leader Éamon de Valera had a clear interest in limiting the flow of volunteers to Spain for both domestic and international reasons. First, the leading figure behind the recruitment of Irishmen to fight alongside Nationalist forces in Spain was General O'Duffy, de Valera's political opponent. Second, complaints were raised in the British House of Commons on 4 February about the fact that O'Duffy's brigade had transited through Liverpool on its way to Spain. Another international consideration was the simultaneous move by several other states to extend their existing non-intervention obligations to include a ban on the entry of foreign volunteers to Spain. The Irish government subsequently confronted fierce opposition in the Dáil (the lower house) and eventually succeeded in introducing the Spanish Civil War (Non-Intervention) Act.[45] Offenders risked a fine or imprisonment of up to two years, or both.[46] However, when O'Duffy and his men returned from Spain, they did not face legal action.

A very different policy was pursued during the Second World War. While it officially remained neutral throughout the conflict, no act was passed in Éire proscribing Irish citizens from enlisting in foreign armies. Tens of thousands of Irish men and women served in the British armed forces between 1939 and 1945.[47] In 1941 the British prime minister, Winston Churchill, inquired whether the oath of allegiance to the king, normally an enlistment requirement, was proving an obstacle to Irish recruitment. However, he was reassured that the oath did not seem to be impeding Irish enlistment as thousands had already taken it and recruitment was continuing apace. In practice, then, the Irish government followed a policy of "benevolent neutrality" toward Britain.[48]

On paper, the United States had more robust legislation than Britain or Ireland. This was because throughout most of the period under discussion, US citizens enlisting to serve another country could lose their US citizenship if an oath of allegiance to the other country was taken. In practice, however, the approach of the US government also exhibited a good deal of flexibility. During the 1820s some 10,000 Americans migrated to Texas. When the Texian colonists, as they were known at the time, rose in revolt against the Mexican government in 1835, volunteers for the Texian cause were recruited in New Orleans, New York, and other parts of the United States.[49] They were promised large tracts of land each, though many of the volunteers were not interested in settling in Texas. In total some 1,500 Americans joined the war in Texas between October 1835 and April 1836. The Mexican government tried to pressure Washington to respect its 1818 Neutrality Act, which did not allow US citizens to interfere in foreign wars, but to no avail. The administration of President Andrew Jackson was pro-Texian, as were authorities in Alabama, Georgia, and other states. The prevailing view was that the Neutrality Act applied "to setting on foot military expeditions to be carried on from the United States against a friendly power." The passage of unorganized volunteers, however, did not violate the law of the land. Hence, the United States turned a blind eye to recruitment for the Texas Revolution.[50]

The costly and, in retrospect, unpopular intervention of the United States in the First World War led to a rise in isolationist sentiments. Against a backdrop of conflicts in other parts of the world—the Italo-Ethiopian War, the Spanish Civil War, the Sino-Japanese War, and, eventually the outbreak of the Second World War—the United States passed a number of neutrality acts between 1935 and 1939. These further restricted the ability of US citizens to volunteer for foreign military service. Moreover, according to Chapter IV of the Nationality Act of 1940, a "person who is a national of the United States, whether by birth or naturalization, shall lose his nationality by: . . . (c) Entering, or serving in, the armed forces of a foreign state unless expressly authorized by the laws of the United States."[51]

There were some signs that the more robust legislation would also lead to greater restrictions in practice. In December 1939 more than a dozen indictments were obtained in Detroit against citizens who had allegedly recruited volunteers for the Spanish Republic. In February 1940 the defendants were arrested by the FBI. However, on assuming office, Attorney General Robert H. Jackson ordered the indictments withdrawn. With the growing plight of the European democracies, the Franklin Roosevelt administration opted to revert to a flexible approach toward foreign enlistment. On 26 January 1940 the president announced that Americans were free to volunteer for service in foreign armies without loss of citizenship so long as they did not take an oath of allegiance to the belligerent power. With regard to the Winter War between the Soviet Union and Finland, he pointed out that "while foreign nations were prohibited from campaigning for enlistments in this country, there was nothing to prevent Americans from inquiring, for example, at the Finnish Legation with regard to service with the Finnish Army."[52]

Loss of citizenship and violation of the neutrality acts were a concern for at least some of the US volunteers who crossed the border into Canada in 1939 and 1940 with the intention of joining either the Royal Canadian Air Force or the British Royal Air Force (RAF). Considering its desperate predicament after the fall of France, the British Air Ministry was very keen to recruit airmen from the United States. Hence, an advertisement was placed in the New York *Herald Tribune* on 15 July 1940, inviting experienced airmen to join the RAF. In October of that year the first of three Eagle Squadrons, consisting almost entirely of American volunteers, was established. In order not to violate US legislation, prospective volunteers were told they would have to enlist in Montreal, Canada, and that they would not need to swear an allegiance to the British Crown. In fact, the British government had agreed to waive the oath for American airmen to facilitate their recruitment. This paved the way for 244 American pilots who eventually joined the Eagle Squadrons. Once the United States entered the war, any fears about legal actions against the volunteers evaporated. In 1942, most of the pilots were incorporated into the US Army Air Force, which valued their wartime experience. Some of the pilots who had joined the war early—before Roosevelt relaxed restrictions—and had temporarily lost their US citizenship, had it restored.[53]

Authorities in the United States responded more firmly toward the involvement of their citizens in the conflict in Palestine in 1948, though here too the vast majority of those who broke the law were not prosecuted. The Neutrality Act of 1939 and the Nationality Act of 1940 were still in force. In addition to these, the State Department expressed concern over the export of weapons and ammunition to the Jews in Palestine and the neighboring Arab states even before the conflict erupted.[54] Thus, the US government imposed a weapons embargo on the region on 5 December 1947. This measure did not prevent American Jews and representatives of the future State of Israel from purchasing weapons and recruiting military personnel. Eventually, surveillance by the FBI and the Treasury Department managed to bring about a small number of arrests. The most notable of these was Charlie Winters, a Boston-born Protestant of Irish-American decent. During the Second World War, Winters worked as a government purchasing agent. He used this experience to obtain three B-17 bombers for the Israelis. He reportedly flew one of them to a training camp that had been set up by the nascent Israeli air force in Czechoslovakia. Winters was imprisoned for eighteen months for having violated the 1939 Neutrality Act and the weapons embargo. He was pardoned posthumously by President George W. Bush in 2008.[55]

Most of the volunteers from the United States who fought for the Israelis during the war of 1948 served in the air force. Initially, steps were taken not to endanger their American citizenship. The Air Transport Command, which flew supplies and war materiel from Europe to Israel, was an all-foreign volunteer operation. Its men were not formally incorporated into the IDF and did not have to take an oath of allegiance to the new state. However, in late 1948, with the war drawing to an end, the leadership of the air force wanted to standardize the service of its personnel and to instill greater discipline among the notoriously unruly foreign airmen. It therefore sought to formally commission the latter as officers. Many of the members of Air Transport Command responded angrily, threatening to go on strike. In a heated debate with air force commander Ahron Remez, the volunteer Harold Livingston pulled out his American passport and recited a paragraph on its last page which stated that nationality may be lost through taking an oath or serving in the military force of a foreign state. Remez refused to be coerced and eventually most of the foreign airmen agreed to the air force's demand. In retrospect, Harold Livingston need not have been so concerned over losing his citizenship. Volunteers from the United States who fought for the Israelis were not stripped of their citizenship.[56] In fact, one of them—David "Mickey"

Marcus, a former colonel in the US Army who died while in IDF service—was buried with full military honors at the Military Academy at West Point.

The democracy that probably has the most robust and least flexible approach toward foreign enlistment is Switzerland, a country that for many centuries was associated with the supply of mercenaries. The Swiss had developed a reputation for being fearless, ruthless, and highly proficient soldiers by the fifteenth century. The cantons recruited men to units that were led by Swiss officers and went on to serve the governments that had signed treaties with the Swiss confederation. Highly decentralized and, initially, economically weak, the confederation established a series of contracts with a number of European states to guarantee for the former political recognition and economic privileges. As we have seen, the demand for mercenaries—including the Swiss—began to shrink as a result of the ideas of the Enlightenment, the rise of nationalism in Europe, and the political and military upheavals of the late eighteenth and early nineteenth centuries. On the supply side, the advent of industrialization meant that labor surplus in Switzerland also shrunk. In the first half of the nineteenth century, the industrializing, Protestant cantons began to press for a single federated system to replace the loose confederation of old. Agrarian Catholic cantons, led by Luzern, resisted, leading to a brief civil war in 1847. The latter's defeat heralded the birth of contemporary Switzerland. The constitution of 1848 prohibited the signing of new mercenary treaties. When the last existing treaty expired in 1859, Swiss mercenary regiments were outlawed. Neutrality became a central facet of Switzerland's foreign policy.[57]

According to the Swiss military code of 1927, military service abroad not only undermined the country's neutrality but also harmed its defensive capabilities (the Swiss military relied on reservists, so the code applied to civilians as well as regular soldiers). Hence, foreign military service without the government's permission was punishable by up to three years' imprisonment or a fine.[58] While the pope was permitted to continue recruiting his Swiss Guards from Catholic cantons, unauthorized foreign volunteers were punished more severely than their contemporaries in other Western countries.

During the Spanish Civil War an estimated 800 volunteers from Switzerland fought on the Republican side. Up to 170 of them may have died during the war. Of those that returned, no fewer than 420 were sentenced to prison terms that varied from two weeks to four years.[59] A Swiss volunteer who had traveled to Finland during the Winter War of 1939–1940 returned to Switzerland and faced court martial in December 1942, even though he did not see combat.[60] The Swiss government also issued warrants for the arrest of known Swiss Waffen SS members in 1944, before the Second World War ended. The authorities revoked the citizenship of the most senior Swiss Waffen SS members and refused entry to Swiss volunteers who were fleeing Allied captivity in the final days of the war. In 1947 several former volunteers were put on trial, and one of them, the high-ranking Franz Riedweg, was sentenced in absentia to sixteen years in prison for treason.[61]

The harshness of Swiss measures against foreign enlistment pales in comparison to those taken by the Soviet Union. For Stalin in particular, the key concern was not preserving his country's neutrality but preventing Soviet citizens from identifying with causes not sanctioned by Moscow. He distrusted ethnic minority groups that could, in theory, have an affinity with and work in the service of other states. His treatment of the Volga Germans, who were deported en masse to Siberia during the Second World War, was one extreme example.[62]

Many Jews in the Soviet Union were swept by a wave of enthusiasm surrounding the birth of Israel in 1948. At first there appeared to be no risk in expressing such enthusiasm as the Soviet Union was favorable toward the creation of a Jewish state and had supported the partition of Palestine at the United Nations in November 1947. In May 1948, after Israeli independence was declared and with the war in the Middle East escalating, some Soviet Jews began to express their willingness to enlist in the Israeli military. Things began to go too far for the Soviet government in September 1948, when Golda Meir, the first Israeli ambassador and future prime minister of Israel, arrived in Moscow. After a crowd numbering in the tens of thousands came to the capital's Great Synagogue on the Jewish new year to see Meir and to express their support for Israel, a crackdown against pro-Zionist manifestations began. Among the many who were arrested, put on trial, and given prison sentences of up to twenty-five years in December 1948 and early 1949 were those who had expressed their interest in volunteering to fight for Israel.[63]

As the examples of Bosnia, Israel, Sweden, Canada, Britain, Ireland, the United States, Switzerland, and the Soviet Union illustrate, the severity of foreign enlistment legislation varies. More importantly, the extent to which such legislation was implemented has nearly always been subject to domestic and foreign policy considerations. Whenever it was opportune to do so, governments turned a blind eye and in some cases even encouraged the foreign enlistment of their citizens. In democratic regimes, the resolve to prosecute foreign volunteers when they returned from their wartime service was often absent. Only in the 2000s, with rising global concerns about terrorism, has prosecuting returning foreign volunteers become a high priority. A dictatorship such as the Soviet Union could exercise far stricter control than most democratic states. Stalin barred Russian Jews from fighting for Israel even though, traditionally, diaspora volunteers have posed little threat to the country in which they reside. However, even Stalin could not prevent the volunteering of substitute-conflict volunteers such as the White Russians who fought for Franco in Spain. A more effective mechanism to prevent foreign volunteers from reaching their destination would require international cooperation. Such cooperation, as we shall see, has never been very forthcoming.

Attempts at International Cooperation

Internationally, much of the customary law relating to foreign volunteers derives from articles 4, 5, and 6 of the 1907 Hague Conventions (V) Respecting the Rights and Duties of Neutral Powers and Persons in Case of War on Land. Article 4 stipulates that "corps of combatants cannot be formed nor recruiting agencies opened on the territory of a neutral Power to assist the belligerents." Article 5 calls upon powers that signed the convention to punish acts that violate their neutrality. Article 6, however, limits the responsibility of neutral states for the actions of individual citizens: "The responsibility of a neutral Power is not engaged by the fact of persons crossing the frontier separately to offer their services to one of the belligerents."[64] In other words, neutral states were called upon to prevent the organization of a hostile expedition but were not expected to prevent individual volunteers from crossing their borders. As we have seen, the duty to prevent organized enlistment has not always been observed. Furthermore, the wording of the Convention has allowed states to disavow any formal involvement in the process and to claim that volunteers were acting independently, as Sweden did in 1940 when faced with Soviet protests over recruitment for Finland.[65]

In the few instances where the international community attempted to further restrict the movement of foreign volunteers in a coordinated manner, the results were very meager. The best-known example of such an international attempt was the Non-Intervention Agreement and the subsequent Non-Intervention Committee during the Spanish Civil War. The agreement came into being in August 1936. It followed a French proposal, and its primary objective was to stem the flow of weapons and war materiel into Spain. To oversee the implementation of this agreement, which several countries joined, a committee was set up in London in September composed of ambassadors and chaired by the Conservative undersecretary to the British Foreign Office, the Earl of Plymouth. The meetings of the committee were dominated by the major European powers. Each of these had its own interests in the conflict in Spain. Fascist Italy and Nazi Germany supported Franco and the Nationalists. In contrast, the Soviet Union supported the Spanish Republic. The Popular Front government in France was favorable toward Republican Spain but faced fierce domestic opposition from the Right, where sentiments were entirely with the Spanish Nationalists. In Britain the Spanish Republic enjoyed substantial popular support but met with a lukewarm response in Westminster.[66] Steered by such conflicting interests, the committee's chances of succeeding were never very high.

In late 1936 and early 1937 the British foreign secretary, Anthony Eden, led an initiative to expand the Non-Intervention Agreement to include a ban on the passage of volunteers to Spain. To give credence to this initiative, the British government issued its declaration of 11 January 1937 that the Foreign Enlistment Act of 1870 applied to the conflict in Spain. Other countries accepted Eden's proposal, and on 16 February the Non-Intervention Agreement was extended. As we have seen, other countries such as Canada, Ireland, and Sweden passed similar legislation around the same time, while France formally closed its border with Spain.[67] Several European states declared their passports to be invalid for travel in Spain. The United States followed suit, even though it was not party to the Non-Intervention Agreement.[68] In spring 1937 the committee set up an elaborate system of naval patrols around Spain's coasts. It also posted more than 100 observers to the non-Spanish side of Spain's frontiers and enlisted hundreds of international observers to sail on ships destined for Spanish ports.[69]

However, a combination of personal determination and networks of support enabled volunteers to bypass the restrictions imposed by the Non-Intervention Committee. The story of Irishman Robert Doyle, a former member of the Irish Republican Army, is a case in point. Doyle was hostile toward O'Duffy and the Blueshirts in Ireland. He tried to reach Spain on his own in various ways. At one point in 1937 he boarded a ship in southern France as a stowaway but was caught after four hours while still en route to Spain. He was disappointed to find a Non-Intervention Officer on board. Released soon afterward, Doyle found a job in a shipping company, sailing to and from Spain. Eventually he managed to make his way to Republican Spain and joined the war in the late summer of 1937 using the network established by the Comintern.[70] The activities of this network in France were managed by the French Communist Party, which found ways to smuggle the volunteers on foot across the Pyrenees. Indeed, as we will see, networks that helped transport volunteers to the seat of war played a key role in overcoming state and international restrictions in a number of conflicts.

The civil war in Syria gave rise to another attempt at international cooperation to stem the flow of foreign volunteers. On 24 September 2014 the UN Security Council called on all member states "to cooperate urgently on preventing the international flow of terrorist fighters to and from conflict zones." In a special summit presided over by US president Barack Obama, the Security Council unanimously adopted Resolution 2178 which expressed

> *grave concern* over the acute and growing threat posed by foreign terrorist fighters, namely individuals who travel to a State other than their States of residence or nationality for the purpose of . . . planning, or preparation of, or participation in, terrorist acts or the providing or receiving of terrorist training, including in connection with armed conflict.[71]

However, lacking any means with which to compel states to comply, all the resolution could do was to urge UN members to "take action, as appropriate, in compliance with their obligations under their domestic law and international law" and to "intensify and accelerate the exchange of operational information regarding actions or movements of terrorists or terrorist networks, including foreign terrorist fighters." It also called on states to take measures that would make it more difficult for suspect individuals to recruit, travel and finance their activities.

A crucial component in any attempt to stem the flow of foreign volunteers into the Syrian conflict hinges on the ability to prevent suspected individuals from entering the country through one of its borders. Much international pressure has been placed on neighboring Turkey to prevent foreign volunteers from crossing its long border with Syria. At the UN summit in September 2014, the Turkish president, Recep Tayyip Erdoğan, said that other countries should do more to prevent such individuals from entering Turkey in the first place and that "cooperation had been insufficient." At the same time he also expressed optimism, pointing out that 3,600 people had already been placed on a no-entry list.[72]

While the available sources do not enable us to form a clear or complete picture of the extent of international intelligence cooperation in this respect, it is worth pointing out that mutual allegations between Turkey and other states continued well after the adoption of Resolution 2178. For instance, on more than one occasion in 2015 the Obama administration criticized Turkey for not monitoring their border with Syria more effectively.[73] Turkey, for its part, also expressed dissatisfaction with the state of international intelligence cooperation. In November 2015 a Turkish official pointed out that his government had provided information about Ismael Omar Mostefai, who had passed through Turkey en route from Syria, to the French authorities on two occasions but did not receive any sign of interest until after the terrorist attacks in Paris.[74] Even if we set aside the quarrel between Turkey and other governments, estimates published by the United Nations itself in 2015 about the growing number of foreign fighters illustrate that Resolution 2178 has done little to remedy the situation. In fact, in April of that year a UN report estimated that the number of foreign fighters worldwide had soared by 71 percent in the preceding nine months, bringing the total that had reached Syria and Iraq to approximately 22,000.[75] Ultimately, the key factors in reducing the number of foreigners who joined and fought for the Islamic State were the military defeats the organization suffered and the territories it had lost in 2016 and 2017.[76]

The success of international cooperation is dependent on the goodwill of the participating states. And states invariably pursue their own interests first. Controlling the flow of foreign volunteers into conflict zones is tied into a broader set of state interests. In the case of the Spanish Civil War, the Soviet Union not only provided weapons and military instructors for the Spanish Republic but also sought to exert influence on the latter's governmental decision making. For their part, Italy and, to a lesser extent, Germany were heavily invested—politically and militarily—in Franco's war effort. Seen from their perspective, the foreign volunteers were pawns in a much larger chess game. Several countries also have conflicting interests in the civil war in Syria. These interests influence their approach toward the foreign fighters in this conflict. For instance, the Turkish government is opposed to the Assad regime and wants to see the Syrian dictator removed from power.[77] At the same time, key concerns for the government in Ankara include the impact of the Syrian conflict on separatist Turkish Kurds and the fate of nearly 3 million refugees that Turkey hosts.[78] Like any other state, Turkey has had to consider the costs and benefits of clamping down on the transit of foreign fighters. Hence, much like in the domestic arena, political expediency plays a crucial role in how foreign volunteers are treated in the states they transit and by the international community.

Borders, Passports, and Networks

Aside from legal action and international cooperation, states have had a limited number of tools with which to prevent foreign volunteering. One of these tools has been the issuing and withholding of passports. In Europe, the requirement of holding passports for the purpose of international travel was boosted by the French Revolutionary and Napoleonic Wars. However, the laissez-faire atmosphere of the nineteenth century saw a relaxation on travel restrictions, and the need for passports was often waived. During the First World War, it was still comparatively easy for volunteers from the United States to take a train across the Canadian border, to pretend to be Canadian, and to enlist. Once in uniform, the costs of passage across the Atlantic were covered.[79] However, the protracted global conflict and its aftermath made the use of passports much more universal. In the interwar years, passports became prized possessions, especially for the millions of stateless refugees that the war and the disintegration of the Russian, Ottoman, and Habsburg empires produced.[80]

As we have seen, during the Spanish Civil War several countries stopped issuing passports to those who intended to travel to Spain or were suspected of planning to do so. However, these restrictions could be circumvented. For instance, the British Communist Party instructed volunteers to purchase three-day-return train tickets to Paris because such a journey did not require passports. Afterward the border between France and Spain was, in any case, crossed illegally, so once again no passport was needed.[81]

The seizure of passports has been used very sparingly as a measure to prevent foreign fighters from traveling to join the civil war in Syria. In February 2015 French authorities confiscated the passports of six citizens who, the intelligence services believed, wanted to join the Islamic State. The measure was part of counterterrorism laws that had been adopted in November 2014. Other European states adopted similar measures. In Britain, for instance, the police were authorized to seize passports for up to thirty days from potential foreign fighters trying to leave the country, and to temporarily prevent nationals suspected of involvement with the Islamic State from re-entering Britain. Twenty-five passports had already been removed from suspected jihadists by February 2015.[82] However, as the vast number of foreign volunteers in both the Spanish and the Syrian civil wars illustrate, the limiting or seizure of passports have not been effective in preventing the departure of committed individuals.

In addition to passports, states have also used their intelligence services and border-control agencies to prevent would-be volunteers from leaving. Sweden, for instance, attempted to do so during the Second World War. In the late 1930s, Åke Kretz was the chief of the air defense division of the Gothenburg police. He was also a reserve lieutenant in the Swedish army. In the spring of 1941 Kretz decided to join the Waffen SS. The Swedish police learned of his intention through a letter sent via diplomatic courier from the German embassy in Stockholm and subsequently intercepted. Kretz came under police surveillance, including a wiretap of his telephone. The wiretaps revealed that he planned to travel to German-occupied Norway to join the Waffen SS. Therefore, his application for permission to travel to Norway—ostensibly to sell his latest book on air defenses—was denied. Kretz resorted to volunteering to join Finnish forces instead. The Finns were engaged in the so-called Continuation War against the Soviet Union, their attempt to undo the results of the Winter War of 1939–1940. Kretz remained on the Finnish front until December 1941. Upon returning to Sweden, he again applied for permission to travel to Norway with the intent of continuing on to Germany, but to no avail. The Swedish police and military repeatedly foiled his attempts at crossing the border into Norway.[83]

But states, especially democratic ones, often find that their ability to thwart the plans of committed volunteers is limited. One of the problems that states faced in the past, and still face today, was anticipating and having enough evidence to prove that an individual intended to travel abroad to become a foreign volunteer. Lacking firm evidence, authorities in Western countries have had to rely on deterrence instead.

During the Spanish Civil War there was little that British police could do to deter individuals wishing to join the International Brigades. A few former volunteers recalled being questioned by plainclothes policemen in Victoria Station or at the ports of Dover, Folkestone, and Newhaven. Asked about the reason for their trip to Paris, the volunteers invariably explained that they were traveling for touristic purposes. All the policemen could do was to issue threats and rely on intimidation. "Well you'd better be back on Monday or else," one volunteer was told in August 1937.[84] This may have worked in individual cases—some are known to have tried to make the journey more than once while others may have given up—but in general this approach failed. During the Second World War there were cases where volunteers from the United States, seeking to cross the border with Canada, were intimidated by the FBI. Agents who approached six volunteers shortly before their train crossed the border gave them a choice: either face arrest or go back home. The six chose to go back home. Soon afterward, however, they tried again and this time managed to reach Canada, and from there Britain, without any interference.[85]

Intimidation was also used on volunteers making their way from Britain to Israel in 1948. Menachem Silberstein, a Holocaust survivor who had found refuge in Britain after the Second World War, was stopped at the port as he was about to leave for Israel in the summer of 1948. The army surplus uniform and boots in his baggage aroused suspicion, and he and those who were traveling with him faced a lengthy questioning. Silberstein ended up admitting that he was en route to Israel. However, the official who questioned him conceded that, because Silberstein's papers were in order, there was nothing he could do to prevent him from leaving.[86]

Volunteers' chances of reaching their destinations were increased by relying on organizations and networks. Some of these were formed in the volunteers' home state, others were set up on a transnational basis, and others still were managed by the host state. In the nineteenth century, networks played a crucial role in ferrying volunteers to war zones that were otherwise too distant and costly to reach. For instance, very few of the laborers and artisans who enlisted in Britain and Ireland to fight for Bolívar could have afforded to make their own way across the Atlantic in the 1810s and early 1820s. The expeditions were financed by traders and investors in London who were hoping to establish positions of advantage for themselves in Gran Colombia. The same people also invested in South American agriculture and mining.[87]

In late 1822 the Greek Society in Darmstadt, today in western Germany, provided weapons, ammunition, and uniforms for a group of 120 volunteers for the war against the Ottomans. Since travel by land to southeastern Europe was difficult and time consuming, the society obtained the brig *Scipio* to ferry the volunteers from the port of Marseilles to Greece. It also provided food, money, and medical supplies for the journey. As we shall see in Chapter 7, this expedition, the first contingent of a planned German Legion, ended in calamity. For the time being, it is worth emphasizing how organized networks were able to provide resources that helped bring individuals, most of whom came from the lower orders of society, to the conflict.[88]

Organizations and networks had another important benefit: they helped volunteers navigate through the legal gray zone. To avoid detection under the Foreign Enlistment Act of 1819, recruiters for the Brazilian Navy maintained the fiction that the recruits were settlers immigrating to Brazil. Thus, in official documents, the seamen that shipped out in early 1823 were described as "labourers" and officers as "overseers."[89] When supporters of Garibaldi in Britain came under criticism for openly recruiting volunteers in violation of the Foreign Enlistment Act, they began to publish advertisements in the press in autumn 1860 calling for "excursionists" who would be willing to travel to southern Italy.[90]

The recruitment and transport networks set up by national communist parties and the Comintern were important in accounting for the large size of the International Brigades during the Spanish Civil War. The success of this endeavor stands out when we consider the scarcity of antifascist volunteers who fought in the Italo-Ethiopian War only a few months earlier.[91] In several countries national communist parties recruited volunteers and arranged their travel to the network's hub in Paris. The breadth and relative efficiency of the Comintern network became the envy of some of their opponents. The Conservative British volunteer Peter Kemp recalled that, "having decided to join the Nationalists I had no idea how to set about it. . . . Had I wished to join the International Brigades there would have been no problem; in every country there were organisations to attract volunteers."[92] The Comintern paid for accommodation in Paris, fed the volunteers while they were there, and transported them to southern France and from there to Spain by boat, bus, or on foot. The importance of these networks becomes self-evident when we consider that the majority of recruits came from a working-class background.[93] An effective foreign volunteer transport network can draw on broader support in the state of transit. This is precisely what the Comintern network was able to do in France. For instance, large groups of volunteers were hosted in and able to pass through the southern city of Perpignan near the Pyrenees, where the cause of the Spanish Republic enjoyed considerable popularity. Furthermore, during the early months of the conflict in 1936, French customs officials and frontier guards were reportedly ordered to turn a blind eye while volunteers crossed into Spain.[94]

Networks were far less important in conflicts where there were few national and international restrictions limiting the volunteers' movement. Moreover, by the 1990s travel had become much cheaper and faster than it had been in earlier periods. The British volunteer Steve Gaunt reached the conflict in the former Yugoslavia in November 1991 without relying on any network. Once he decided to volunteer, he bought tickets to fly to Venice and was in Zagreb five days later.[95] Fellow British volunteer Simon Hutt bought a one-way flight ticket from Gatwick to Vienna and from there caught a train to Zagreb.[96]

For the civil war in Syria much of the recruiting network exists online, where advice on travel arrangements can also be obtained.[97] Compared to their nineteenth- and twentieth-century predecessors, online recruiters are able to cast a much wider net in terms of reaching potential volunteers. Moreover, if the recruiters are able to carry out their work from Syria, it is much more difficult for Western governments to apprehend them. Most of the foreign volunteers have reached the war zone through Turkey. Travel to Turkey, certainly from most European countries, is relatively cheap and does not require an entry visa. A contemporary volunteer would usually come face to face with the network for the first time in order to cross the border from Turkey into Syria. Local smugglers are used for this purpose.[98]

As networks funneled volunteers to conflict zones, they also exercised the ability to prevent passage. During the Spanish Civil War national communist parties and the Comintern refused to enlist individuals they deemed undesirable in the International Brigades. British volunteer Jason Gurney recalled going to the Communist Party office on King Street in London so that he could join the war in Spain. He and a number of other eager young men received a stern lecture from a party official called R. W. Robson about the difficulties they were likely to face: "It was a bastard war, we would be short of food, medical services and even arms and ammunition." Those who believed they were going into a fine adventure were advised to go home straight away. When one of the would-be volunteers insisted about the conditions of service, Robson snapped: "If you're looking for conditions of service, you're not the kind of bloke we want in Spain. So get out."[99]

Applicants were also turned down for political reasons. When Eric Blair (better known by his pen name, George Orwell) made inquiries about joining the International Brigades, he was interviewed by Harry Pollitt, the general secretary of the Communist Party of Great Britain. The two men had a bad-tempered conversation. Pollitt refused to accept Blair into the Brigades because he was politically unreliable. The author subsequently made his own way to Spain.[100] The French Communist Party carried out further screening, including medical examinations, for the volunteers who passed through France, though many volunteers were able to evade these checks.[101]

In 1948 the Israelis received more offers to volunteer than they could handle. This placed them in the privileged position of being able to take the foreign volunteers they wanted and turn down the rest. The new state's diplomatic representatives in Rome received dozens of letters from individuals prepared to offer their services to fight for Israel's independence. For instance, Emilio Benvenuto from Foggia wrote in June to express his "admiration for the heroism of the Sionist fighting men." As an Italian Catholic, he felt compelled "to pay the debt that I have towards the people of the Fathers of my religion and to pay also the debt that my country has contracted towards the Judish [sic] people with the past adoption of the idiot racist doctrine." He therefore offered "my arm and my life, if useful."[102] Nino Cuffao similarly spoke of how the Nazis and Fascists had persecuted the Jews. A former partisan commander, he offered his services to the Israelis because of the great sympathy he had for the new state. He emphasized that his motives were not material and that life on his farm was, in fact, very good.[103]

In the haphazard conditions of 1948 it took a long time for the Israeli representatives to reply. It was not before the autumn of that year that laconic responses were sent out, explaining that "the Israeli army does not accept volunteers of other nationality" or that the Israeli delegation was "not authorized to handle volunteers."[104] While the IDF did, of course, recruit foreigners, including non-Jews, it did not lack manpower per se. Pilots were sorely needed, and these were recruited wherever they could be found. More generally, however, once the tide of war started to turn in favor of the Israelis, they began to view the Jewish foreign volunteers as potential immigrants and to invest efforts in convincing them to stay.[105] With the country's resources already stretched, and with no immediate military necessity, there was no need to recruit non-Jewish volunteers, even if their motives were idealistic.

In addition to controlling access to the war zone, recruitment and transport networks also sought to monitor the foreign volunteers' movement. Like the volunteers' home states, the networks recognized the value and importance of passports. In Paris in 1937 a bilingual *"responsable"* was appointed by each of the main communist parties to look after its national contingent of volunteers during their stay in the French capital. While solving various logistical problems, these representatives were also told to demand each volunteer to surrender his passport. Charlotte Haldane, the British *responsable,* later explained that this idea came from the hosting French Communist Party, which sought to protect itself in case a foreign man was arrested by the local police. The Italian and German representatives concurred, hoping—in case of an arrest—to protect their underground antifascist movements. The volunteers, however, resented this demand. Haldane observed, "Those of them who did have legal papers regarded them as their most precious possessions."[106] Such concerns were vindicated when the International Brigades were disbanded in 1938. It proved difficult to extract British and American volunteers without their passports while German, Italian and other "stateless" men found themselves in French detention camps.

The network that funneled foreign volunteers through France to Israel also took away the volunteers' passports and provided them instead with false identity papers. The volunteers were not told why such a move was necessary. One explanation that has been put forward is that the Israelis wanted to mix the volunteers with the displaced Holocaust survivors that were being transported through French and Italian ports to Israel as immigrants.[107] By doing so, the transport network was hoping to evade a ban imposed on 29 May 1948 by the UN Security Council on the introduction of "fighting personnel" into the warring countries.[108] As a result of having their passports taken away, those volunteers who changed their mind about fighting in the Middle East while waiting in a camp in southern France had to appeal to their home state's diplomatic representative in Marseilles to facilitate their return journey.[109] Passports were returned to those volunteers who were demobilized at the end of their service.[110]

Some of the foreigners who joined the ranks of the Islamic State in Syria also had their passports taken away. Ahmed Shurbaji, for instance, was told to surrender his Israeli passport in 2014. As with previous generations of foreign volunteers, the absence of identification documents restricted his ability to leave the conflict zone.[111] It can be assumed that keeping him in Syria was one of the intentions of his recruiters. The Islamic State has reportedly executed foreign volunteers who expressed the desire to leave their ranks.[112] Whether the organization (or previous networks, for that matter) wanted to make further illicit use of such coveted commodities as foreign passports is a matter of speculation.

※　※　※

In conclusion, while de jure there was, and still is, an international norm of preventing the departure of foreign volunteers, de facto the situation has been, and remains, much more complex. Several states took measures to limit foreign volunteering to show the international community as well as their domestic public that the issue was being addressed. However, considering all the vying interests, the result has often been attempts to bring the flow of foreign volunteers under control rather than seriously preventing it.

Furthermore, in some cases governments were very sympathetic toward citizens who wished to serve causes that were perceived as justified, as the Swedish policy regarding volunteering for Finland in 1939–1940 illustrates. There is also some evidence to suggest that, on a few rare occasions, some governments saw the outbreak of an ideologically charged conflict abroad as an opportunity to get rid of troublesome extremists. A number of Jewish communists who left Palestine to fight in the Spanish Civil War testified that the British Mandate police had released them from prison on the understanding that they would go to Spain.[113] According to Milton Bearden, who served as the CIA's station chief in Pakistan between 1986 and 1989, "a number of Arab states discreetly emptied their prisons of homegrown troublemakers and sent them off to the jihad [in Afghanistan] with the fervent hope that they might not return."[114] In other words, under certain circumstances, the departure of foreign volunteers was seen as a blessing rather than a problem.

In the 1820s the state was smaller than it is today and had far less information about its population. In theory, this made foreign volunteering easier, but travel was much more costly and time consuming. In the 2010s more intrusive governments use the latest technology in an attempt to catch up with the opportunities created by cheap and fast travel. In October 2014, for instance, the FBI managed to intercept three Syria-bound teenage girls from Denver in Frankfurt, Germany. They had reportedly stolen some money from their parents to pay for their flights and were apprehended because the FBI flagged their passports.[115] Governments sometimes win in this competition between transport and surveillance technology, but sometimes, from their perspective, they also lose. One is reminded of the apprehension that gripped many in Britain in the 1930s about the dangers and destruction that a future air war would bring. The Conservative leader and future prime minister Stanley Baldwin famously warned in 1932 that, whatever precautions were taken, "the bomber will always get through."[116] The same can be said of committed foreign volunteers who are prepared to try time and again to reach their destination. This raises the question, which the next two chapters address: is the threat they pose really so great?

* 6 *

Winning Wars?

Assessing Military Significance

In September 2014 UN Security Council Resolution 2178 expressed concern that what it called foreign terrorist fighters "increase the intensity, duration and intractability of conflicts."[1] This is a curious assertion, not least because the existing historiography has yet to explore the distinct military impact of previous cohorts of foreign volunteers on a comprehensive, multi-conflict level.[2] This chapter, therefore, turns to military history in an attempt to gauge the wartime significance of such transnational fighters.

Let us begin our analysis by posing a question: do foreign volunteers tend to constitute an effective fighting force? Allan Millett and Williamson Murray developed an approach for the analysis of military effectiveness in the late 1980s, but their focus was on national military forces in, essentially, state-on-state warfare. They understood military effectiveness as "the process by which armed forces convert resources into fighting power."[3] In order to assess the performance of different military organizations in different historical contexts, they devised a list of questions that measured each country's political, strategic, operational, and tactical effectiveness. Most of these state-centered questions do not apply to the transnational contingents that concern us here. However, there are exceptions. The most notable of these is one of the questions that measures political effectiveness: "To what extent do military organizations have access to manpower in the required quantity and quality?" In this context, Millett and Murray did well to point out that access to manpower involves not only legal power but also moral and political legitimacy, an attribute that often appeals to foreign volunteers.[4] Although the three volumes they edited do not examine whether and how military organizations recruited transnationally, the question they posed is a good starting point for a discussion of the military importance of foreign volunteers.

Writing more than two decades later, David Malet put forward a hypothesis that can be used to answer Millett and Murray's question about access to manpower. Focusing strictly on civil wars and insurgencies, Malet argues that "local insurgents, who always begin conflicts as the weaker faction because they do not control the instruments of the state, attempt to broaden the scope of conflict to increase their resources and maximize their chances of victory" by trying to recruit fighters from abroad.[5] He goes on to claim that "transnational insurgents win civil conflicts at a rate disproportionate to other insurgencies."[6] In other words, insurgents who succeed in recruiting foreign volunteers are more likely to win the conflicts they engage in than insurgents who only recruit locally. Malet does not go as far as arguing that such foreigners necessarily tip the military balance, but he does suggest that insurgents who are better organized and more effective have the capacity to recruit abroad.[7]

Alas, there are a number of problems with the assertion that foreign volunteers are more likely to be found on the winning side. The more closely the historical record is analyzed, the more it becomes apparent that foreigners often joined more than one party to each conflict. During the nineteenth-century wars of the Italian Risorgimento, some foreign volunteers joined the various Garibaldian military expeditions while others fought alongside the Bourbons or defended the Papal States.[8] In the Spanish Civil War, foreigners fought for the Republicans and for the Nationalists. Incidentally, in this conflict—as in many others—the initiative to enlist came from abroad rather than from the government or the insurgents. In 1948 the newly established State of Israel recruited foreign volunteers from Europe, South Africa, and North and South America. At the same time, volunteers from Syria, Iraq, Egypt, Lebanon, and other countries fought alongside the Palestinian Arabs. In the 1990s, as we have seen, foreigners who participated in the Yugoslav Wars joined Bosnian, Croatian, and Serb units. During the siege of Kobani and other battles in northern Syria in 2014–2015, some foreigners fought for the Islamic State while others joined the opposing Kurdish forces. Clearly, then, the mere incorporation of foreigners into a fighting force does not tell us much about that side's effectiveness or chances of success.

A clearer picture emerges when we consider the size of foreign contingents within the overall number of belligerents in each conflict. Even in cases where comparatively large numbers of foreign volunteers participated in conflicts abroad, they still constituted a tiny minority. The total number of foreigners in the International Brigades throughout their participation in the Spanish Civil War has been estimated at around 32,000, with a few thousand more serving in other military units on the Republican side or as doctors, nurses, and engineers.[9] While the overall size of the Republican military force is disputed (some estimates put the total number mobilized throughout the war as high as 1.7 million), historians believe that the International Brigades comprised between 2 and 6 percent of all the men and women under arms.[10] During the Winter War of 1939–1940 approximately 12,000 foreign volunteers enlisted to fight for Finland against the Soviet Union. Again, there is some uncertainty about the exact number—for instance, some volunteers were still in transit when the war ended—but the foreigners only constituted about 3.5 percent of the Finnish field army, the strength of which was approximately 346,000 at its peak in March 1940.[11] In 1948 approximately 3,500 foreign volunteers fought on the Israeli side. Their percentage among all the men and women the IDF was able to recruit by the end of that conflict—more than 108,000—was similar to that raised by the Finns eight years earlier.[12]

As we have seen, estimates about the number of foreign volunteers who fought in Afghanistan against the Soviet Union and their local allies vary between 5,000 and 25,000. Abdullah Azzam put the number of foreign volunteers at between 7,500 and 8,500.[13] Meanwhile, the Soviets reckoned that the overall strength of the Afghan mujahideen was between 170,000 and 200,000.[14] Estimates about the number of foreign fighters in the insurgency in Iraq in the early 2000s changed over time. In 2003 and 2004 there were only a few hundred, with numbers rising to up to 2,000 in 2006. Meanwhile, estimates of the total number of insurgents rose from around 5,000 to 20,000 over the same period. Indeed, despite the variation in figures, there seems to be a consensus that foreigners never constituted more than 10 percent in the so-called Sunni insurgency in Iraq.[15] In fact, in all of the conflicts mentioned above, less than one in ten of the soldiers and fighters were foreign while the overwhelming majority were homegrown. If foreign volunteers have been militarily significant, their contribution must be sought not in numbers but in other spheres.

Tactical Effectiveness in Large-Scale European Conflicts

Military historians tend to use a number of indicators to measure the effectiveness of forces and units. These include the extent and rigor of the soldiers' training, efficiency of logistical support, levels of discipline, unit cohesion, and relations between officers, noncommissioned officers, and the enlisted ranks.[16] The ultimate test to a unit's tactical effectiveness has always been how well it conducted itself in battle. Casualty rates, especially the ratio of dead and wounded to prisoners of war, can be used—where such data is available— as a key indicator of the unit's commitment to the fight.[17] Let us therefore consider a few examples of foreign contingents that have acquired an aura of either success or bravery. These are drawn from four large-scale, essentially conventional conflicts: the Franco-Prussian War of 1870–1871, the First World War, the Spanish Civil War, and the Second World War (smaller-scale warfare will be discussed later on).

Various foreign volunteers offered their services to France during the Franco-Prussian War: men (and a few women) from Argentina, Belgium, Brazil, Britain, Greece, Spain, Switzerland, the United States, and Uruguay, along with Polish exiles and, of course, Italians. Giuseppe Garibaldi, though already in his sixties and suffering from ill health, arrived in France in early October 1870, about a month after the fall of Napoleon III and the establishment of the Third Republic. He offered his services to the Government of National Defense, which had moved to Tours to escape the German siege of Paris. The French leadership attempted to mobilize more manpower to resist the invading armies of General Helmuth von Moltke the Elder that now occupied about one-third of France's territory. Although he was received in Tours with some reservation, Garibaldi was put in command of a force of *francs-tireurs* and *mobiles*—French volunteers who came forth to defend the republic rather than regular soldiers of the French army—as well as foreign volunteers, and dispatched to the front near Dijon. With time, this force, known as the Army of the Vosges, grew in size from around 6,000 in November 1870 to over 16,000 by February 1871.[18]

Appraisals of the performance of Garibaldi and his Army of the Vosges varied.[19] Dijon had already capitulated to the Germans on 31 October 1870, and Garibaldi tried to recapture it, with a relatively small force under the cover of darkness, on the night of 25–26 November. However, his attack was comprehensively defeated, and he was forced to retreat. German forces evacuated Dijon in late December, and Garibaldi's troops, after some delays, were able to secure the city by mid-January 1871. They then managed to repel two enemy attacks on Dijon between 21 and 23 January. The German attackers lost 700 men, and Garibaldi's troops managed to capture an enemy flag—a great boost for French morale. Ironically, the armistice of 26 January 1871 tarnished Garibaldi's record because it excluded the ongoing fighting in eastern France. This allowed the German army to amass around Dijon, using forces freed up from combat elsewhere. Garibaldi decided to abandon the city in early February, letting it fall without a fight. Conservative French critics of the aging general highlighted this last episode as well as other cases of insubordination where Garibaldi refused or failed to assist fellow French commanders who fought in the area. Meanwhile, supporters of Garibaldi pointed out that the Army of the Vosges was constantly outnumbered and outgunned by the enemy, and that the *francs-tireurs* displayed great bravery.[20]

It is difficult to assess the performance of the foreign volunteers in the Army of the Vosges for two main reasons. First, much of the historiography has focused on the conduct of Garibaldi himself and his strained relations with the other French commanders (the staunch Catholics among them loathed the Italian hero who was widely perceived as the enemy of the pope). Second, the foreigners, who were grouped into a number of companies in the army's four brigades, were only a minority within a predominantly French force. Garibaldi's troops were effective when they conducted guerrilla-style surprise attacks against enemy patrols and positions, operating in small parties and moving under the cover of darkness. Ricciotti Garibaldi, one of Giuseppe's sons and commander of the army's 4th Brigade, carried out a successful surprise raid on Châtillon-sur-Seine in mid-November. However, Garibaldi's troops were far less effective when it came to pitched battles, as their unsuccessful attack on Dijon later that month illustrated. Indeed, their daring assault was no match for the Germans' firepower and discipline. Soldiers in the Army of the Vosges were committed—in one battalion the casualty-to-prisoners-of-war ratio was 7:3—but it is unclear whether and how foreign volunteers compared to French troops in that respect.[21] Overall, an estimated 3,000 Italians fought alongside the French in the Franco-Prussian War, mainly with Garibaldi but also in other units. According to one contemporary report, there were still 2,865 Italian volunteers in French service in March 1871. Just over 100 Italian volunteers died during the war.[22] The percentage of deaths among the men that enlisted, while sizable, was smaller than that suffered by some of the other contingents that fought in the Franco-Prussian War.[23] Moreover, the figure pales in comparison to those that foreign contingents suffered in the era of total war during the first half of the twentieth century.

The Garibaldian volunteers of 1870–1871 were emulated by another generation of Italians who fought for France in the early months of the First World War, while Italy was still neutral. Some 2,354 volunteers, almost all of whom were Italian, joined the 4th Régiment de Marche of the French Foreign Legion in September 1914. The unit was soon nicknamed the Garibaldi Legion. In a gesture that was unique in the history of the Foreign Legion, French authorities allowed half of the military officers to be Italians, and command of the unit was divided between Peppino Garibaldi (Giuseppe Garibaldi's grandson) and a French colonel.[24] The spectrum of ages in the Garibaldi Legion was very broad. The youngest to enlist was a 14-year-old from Florence who was in Marseilles with his parents when the war broke out. Having run away to join the army, he passed himself off as being 17. Meanwhile, the oldest volunteer, Sergeant Mori, was 60 years old. As a teenager, Mori had joined Garibaldi's campaign in France in 1870–1871.[25]

Despite the heterogeneous background of the volunteers and their cavalier approach toward the Foreign Legion's discipline, the regiment displayed great courage in its first baptism of fire. Posted to the front near Argonne in December 1914, the volunteers saw intense action over a two-week period between late December 1914 and early January 1915. One-third of the soldiers, half of the noncommissioned officers, and 41 percent of the officers were either killed or injured. Many of the dead were decimated by machine-gun fire during attacks that attempted to take German trenches. Among them were two of Giuseppe Garibaldi's grandsons: Bruno and Costante. The Italians' casualty rates were roughly similar to those of French frontline units in late 1914 and 1915. This suggests that their commitment to the cause was strong. However, their experience on the battlefield irrevocably destroyed the unit's cohesion. Following their ordeal, the Garibaldi Legion was transferred to the rear on 10 January 1915. They were posted to Avignon, where discipline broke down completely. The commanders seem to have lost control over their men, some of whom reportedly discharged their weapons from the roof of the Palais des Papes. On 5 March 1915 the French minister of war ordered the Legion's disbandment.[26] The military involvement of Italian volunteers in France in 1870–1871 and again in 1914–1915, therefore, was relatively brief. How did foreign units fare in longer periods of military activity such as in the Spanish Civil War and in the Second World War?

The effectiveness of the International Brigades during the Spanish Civil War has long been a contested subject. The age spectrum of foreign volunteers was larger than that of Spaniards conscripted by the republic. The Madrid government mobilized men between the ages of 18 and 44. Meanwhile, there were foreign volunteers as young as 17 and as old as 56, although the majority were in their twenties and early thirties.[27] Judging by their casualty rates, the Brigades were composed of highly committed individuals. In the fighting around Madrid in January 1937, the predominantly German Thaelmann Battalion refused to retreat under fire and lost two entire companies. Overall the 11th International Brigade lost around 900 killed and wounded on that front alone. The British Battalion of the 15th International Brigade had 275 casualties out of 400 men in its first day of action at the Battle of Jarama on 2 February 1937. Indeed, casualties among the other battalions of the 11th, 12th, and 14th International Brigades that fought at Jarama often exceeded 50 percent. During the Battle of Brunete in July 1937, the British Battalion lost 268 men out of 300 effective in less than three weeks of fighting, and their counterparts from the United States had 620 out of 900 men killed or injured. This battle reduced the 15th International Brigade from a strength of 2,500 men to under 600 effective. Overall, the casualty rates of the International Brigades were higher than those of local and foreign units fighting on the Nationalist side. They also outstripped Allied losses in the Second and even the First World War.[28]

The 11th and 12th International Brigades were especially celebrated, both during the war and in the partisan historiography, for saving Madrid when it came under Nationalist attack in November 1936. Foreign volunteers fought heroically and bore the brunt of Nationalist attacks in the capital's

University City. They were also credited with providing a model for inexperienced Spanish troops. "People learn from example more quickly than from lectures," the British volunteer Tom Wintringham explained in his memoir. "We have to show the Spanish militia what a real army looks like . . . and when they have learnt what we can teach them they will beat Franco."[29] However, in recent decades, dispassionate and revisionist historians have called into question the centrality of the Brigades in the Battle of Madrid as well as the military effectiveness of these foreign volunteers more generally.

To begin with, many of the foreign volunteers had either inadequate or no previous military experience. Some, like the Swedes and Swiss, had served in their national military forces, but others only received rudimentary training at school or university or none at all. Of the Canadian volunteers, for instance, only about 13 percent had previous military experience. The training the volunteers received upon arrival to their camp in Albacete was often minimal. The men of the 11th and 12th International Brigades who were rushed to Madrid in November 1936 often went into action only having fired their rifles once or twice. The training that new recruits received was often focused on parades and political lectures. It did not place any emphasis on the vital skill of digging in. This, among many other reasons, accounted for the huge losses at Jarama and Brunete.[30]

Revisionist historians have been very critical of the commanders of the International Brigades. The generals who led each of the Brigades were appointed by the Comintern. These were often former First World War prisoners of war from Eastern Europe who had joined the Red Army during the Russian Civil War. Assigned to Spain, they were given false names to hide their identity. For instance, Emilio Kléber, who commanded 11th Brigade, was actually Moshe (Manfred) Zalmanovich Stern from Bukovina. General Walter, who led 14th Brigade, was in fact the Pole Karol Świerczewski. General Gal, the commander of 15th Brigade, was the Hungarian János Gálicz. These men were chosen over more experienced officers born within the borders of the Soviet Union, so the government in Moscow would be able to disavow any connection to them. Fully aware of the fate that might meet them back in the Soviet Union for deviating from Stalin's directives, these generals committed the Brigades to battles they could not win and refused to allow them to retreat.[31] Revisionist historians also point out that, in choosing many of the battalion commanders that were appointed to each of the "national" contingents, loyalty and propagandistic value were more important than military acumen. Relative successes of the Brigades, such as in the fighting around Madrid, were ascribed to Nationalist weakness, the effect of other republican units and the intervention of Soviet tanks and aircraft. The Brigades saw themselves as the republic's shock troops although, behind their backs, some government officials reportedly saw them as cannon fodder.[32]

Another problem with assessing the foreign volunteers' effectiveness in Spain is the fact that the International Brigades were only partly international. Spanish volunteers were incorporated into the Brigades starting from December 1936. Their proportion grew over time and, because of the tremendous losses the Brigades suffered, Spanish conscripts also filled the ranks as of April 1937. By December 1937 more than half the soldiers in the Brigades were Spanish. In fact, in early 1938 General Walter estimated that in four out of five of the Brigades, foreigners only constituted some 20 to 30 percent.[33]

The historiography of non-German Waffen SS units and other foreign formations that fought alongside Nazi Germany is similarly contested. Veteran German officers depicted the Waffen SS as an elite force that fostered trans-European camaraderie in the struggle against Bolshevism.[34] Some non-academic Western historians seem to have been taken by the heroic aura that was constructed around the Waffen SS and see them as deserving rehabilitation. Their work emphasizes the courageous efforts of the volunteers.[35] Meanwhile, academic historians have tended to be far more critical of Germany's foreign volunteers, pointing out their involvement in war crimes.[36]

Because of the biases in the debate, assessing the military effectiveness of pro-German foreign contingents is not straightforward. Take for instance the volunteers from France. The Legion des Volontaires Français contre le Bolchevisme (LVF) was established in summer 1941, following the German invasion of the Soviet Union, at the behest of collaborationist leaders in France. Senior officials in the Vichy France government had serious reservations about this initiative. The Germans too were unenthusiastic, placing restrictions and limitations on the size of the French force. Nonetheless, the Legion was incorporated, as an infantry regiment, into the Wehrmacht's 8th Corps in November 1941.[37] Although part of the German Moscow offensive, the regiment was stopped forty miles from the Russian capital, ruled unfit for frontline combat by the German command, and assigned to fighting partisans in Belorussia. Suppressing local resistance remained their role until their retreat from the eastern front in 1944. Remnants of the LVF were then merged into the newly created Waffen SS Charlemagne Division, which also included veterans of the Frankreich Brigade, an earlier Waffen SS outfit that was recruited in France in 1943. Some 7,000 Frenchmen of Charlemagne fought against the Red Army in Pomerania in early 1945, suffering very heavy losses. A few members of the division later participated in the final attempts to defend Berlin in April-May of that year.[38] In his relatively favorable account, Robert Forbes seems to accept the position expressed after the war by Gustav Krukenberg, the former commander of Charlemagne: "Without the Frenchmen, the Russians would have taken Berlin eight days sooner."[39] In contrast, more critical studies have pointed out that the LVF had the worst combat performance record among the non-German units and that, in April 1945, about 50 percent of the 1,200 surviving volunteers of Charlemagne chose to opt out of the fighting in Berlin when given the opportunity to do so.[40]

The record of other foreign units that fought alongside the Germans is less contested. The Freikorps Danmark, for instance, performed relatively well in its engagements with the Red Army in 1942. The Wiking Division, which incorporated Dutch, Norwegian, Danish, Finnish, and Flemish volunteers, was considered one of the best formations in the Waffen SS. However, foreign volunteers constituted a minority in this division. In June 1941 only 1,564 of its 19,377 men were foreigners.[41] In general, Waffen SS units suffered at least 5 percent more fatalities than those of the Wehrmacht. Such higher casualty rates have been attributed to the aggressive and combat-seeking ethos of the SS as well as to the arrogance and at times misplaced confidence of Waffen SS commanders.[42]

But the zeal of the Waffen SS must not be overstated. As we have seen, in the later stages of the war Himmler began to recruit in eastern and southeastern Europe in an attempt to address manpower shortages. The ideological commitment of some of these new national formations was partial at best. The SS Handžar Division (*Handschar*, in German sources), named after the traditional Turkish scimitar, created in February 1943, and composed of Bosnian-Muslim recruits, provides a useful example. Although the local population had hoped that the division would be used to protect their villages and towns, the men of the Handžar were sent for a lengthy period of training in France and Germany. In September 1943 a group of Bosnian and Croat soldiers mutinied at their camp near Toulouse. Once it was sent back to the Balkans, more than 2,000 of the division's members deserted in the autumn of 1944, and the number of self-mutilations rose. The unit was subsequently disbanded and remaining soldiers were given a choice between labor service and joining other SS or Croatian Ustaše outfits. Other Waffen SS formations that were raised in the Balkans, such as the Albanian Skanderbeg Division, also suffered from very high rates of desertion.[43] The foreigners who fought for Germany during the Second World War, much like the foreign contingents who took part in earlier conflicts, were far from being a homogenous cohort, and any generalization about their effectiveness is hazardous.

Linguistic Obstacles and ad hoc Solutions

As the examples above illustrate, whenever foreign volunteers took part in regular warfare, their effectiveness depended on the same variables that military historians always stress: manner of deployment, command, equipment, training, morale, and so on. One disadvantage that foreign volunteers in different periods and different conflicts very often shared was the difficulty to communicate with their hosts.

In his famous novel on the Spanish Civil War, *For Whom the Bell Tolls,* Ernest Hemingway found a literary solution that enabled his protagonist, Robert Jordan, to communicate with the book's Spanish characters: before the war Jordan had taught Spanish at an American university and therefore spoke the language fluently. Moreover, he had spent long periods of time in Spain and had traveled through the country in previous years. Hence, Jordan "never felt like a foreigner in Spanish and they did not really treat him like a foreigner most of the time."[44] However, this level of proficiency was extremely uncommon among foreign volunteers, not only in the Spanish Civil War but in most other conflicts too.

Examples of how the military effectiveness of foreigners could get lost in translation are plentiful. Volunteers in Napoleon's Irish Legion were obliged to learn French at the village of Morlaix, where their training took place. However, as far as the French authorities were concerned, the results were unsatisfactory. In October 1804, after months of training, only 45 percent of the Irish officers for whom data is available could read and write in French at an adequate level.[45] During the First World War Italian officers serving in the 4th Régiment de Marche of the French Foreign Legion had to give orders in French even though they did not necessarily speak the language very well. Meanwhile, the rank and file, who often spoke different Italian dialects, struggled to understand.[46] The Babel of languages in the International Brigades of the Spanish Civil War era made linguistic difficulties inescapable. As Jason Gurney recalled:

> Theoretically the common language of the Brigade was French, a language which none of the Brigade staff nor any of the battalion commanders spoke with any real fluency, except for those of the Franco-Belge Battalion. To make matters worse, the bad French spoken by a Russian or a German is quite unlike the bad French spoken by an Englishman or an Italian. People could just about make themselves understood in normal life but in the heat of battle, over an inadequate telephone line, there was virtually no communication at all.[47]

In Israel during the war of 1948 the number of Anglophone volunteers in the air force was so high that English became the operational language.[48] However, English was of no use to David "Mickey" Marcus, a former US Army colonel of Jewish descent, on the night of 10 June 1948. Put in command of the Jerusalem front, he was accidentally shot dead by a Jewish sentry who called in Hebrew for the password but got no reply.[49] Linguistic difficulties were also encountered by Arabs in Afghanistan in the 1980s and British volunteers in Croatia in the 1990s as well practically every other contingent of foreign war volunteers.[50]

Linguistic considerations as well as cultural preferences often led to the establishment of separate units for foreign personnel. In some cases, the hosting military authorities created such units, while in others the initiative came from the volunteers. In the Italian Risorgimento, for instance, foreigners were integrated into special units by all the sides that took part in its conflicts. The papal army often grouped together soldiers speaking the same language— such as Austrians, Germans, and Swiss Germans—in the same regiments, as far as possible. Meanwhile, the supporters of Italian unification organized

Hungarian and Polish Legions in 1848 and an 800-strong British Legion in 1860.[51]

Foreigners who fought alongside the Greeks in the Greco-Turkish War of 1897 were incorporated into a number of distinct units. The Philhellenic Legion was set up by the Greek government, which wanted to have more control over the foreigners' movements. Ricciotti Garibaldi, who had fought alongside his father in France in 1870, led a separate Red Shirt army composed of 1,323 volunteers and divided into four battalions (two Italian, one Greek, and one "mixed," incorporating various nationalities). Enrico Bertet, a former colonel in the Italian army, and the anarchist Amilcare Cipriani tried to raise smaller separate units, though with less success.[52] In the Spanish Civil War the International Brigades were subdivided into national, or at least linguistically distinct, battalions and companies. In Finland during the Winter War Swedish volunteers served in a separate unit that also included some Norwegians. Everyone else was assigned to another detachment that was subdivided into American, Hungarian, British, and mixed companies.[53] As we have seen, the Second World War Waffen SS, too, established separate divisions or regiments for recruits from different countries. "National" units, even in cases where they integrated volunteers from a number of nationalities, were often led by foreign rather than local officers. This was supposed to ease communication, at least within units, though in practice linguistic difficulties remained a near-constant feature of transnational military service.

Tactical Effectiveness in Improvised Military Formations

Because traditional measurements of battle effectiveness tend to focus on state military forces in regular warfare, they are often ill-suited for assessing the contribution of foreign volunteers. The latter were, and are still, far more likely to join militias, paramilitary units, guerrilla outfits, or newly constituted military forces. In contemporary parlance, the type of fighting that many foreigners engaged in could be called hybrid warfare, a mixture of conventional and irregular practices, with varying degrees of centralized command and planning.[54] Even the foreign volunteers who ended up participating in conventional warfare—such as those who fought in the Franco-Prussian War or in the Spanish Civil War—initially joined hastily organized units, established under post-revolutionary or civil war conditions. In the haphazard conditions that normally prevailed in such military formations the level of record keeping was very poor. Hence, statistical analyses of casualty rates are either not very revealing or simply not possible. Some impressionistic and anecdotal evidence can be drawn from the accounts of foreign volunteers or their local commanders about the former's wartime commitment and effectiveness (or lack thereof). Furthermore, in some cases comparisons can be drawn between the foreigners' performance and that of their hosts. Alas, the picture that emerges from these sources is incomplete and often contradictory. As the following examples illustrate, foreign volunteers could be exemplary soldiers or a burden on their hosts or just about anywhere on the spectrum between these two extremes.

Only about one in three of the foreigners who enlisted to fight for the independence of Gran Colombia from Spain in the 1810s and 1820s had previous military experience from service in the Napoleonic Wars. In some cases, foreigners made extraordinary sacrifices and provided a source of inspiration for local troops. For instance, when the rebel-held fort of Barcelona was about to fall in 1816, the former British armed forces officer Charles Chamberlain killed himself rather than fall into Loyalist hands. Another foreigner, William Ferguson, was killed protecting Bolívar from an assassination attempt. Coronel James Rooke, who died after the Battle of Pantano de Vargas in 1819, became a symbol of the new republican patriotism when the dying words he uttered were *"viva la patria."* Furthermore, veterans of the Royal Navy who joined Bolívar's naval force brought with them a high level of technical experience as well as some of the ethos of confidence and aggression they had acquired in their previous service.[55]

But such heroic and professional individuals were the exception rather than the rule. Arriving in a few contingents over the course of 1816–1820, and suffering from very high attrition rates even before they reached the seats of war, the foreigners as a whole did not bring about an immediate professionalization of Bolívar's army. Historian Matthew Brown calculated that more than half of these foreigners—3,633 out of approximately 6,800—died shortly after arriving in the Caribbean, deserted, or swiftly returned to Europe, having become disenchanted with the cause. In total, about 22 percent died of fever and some 26 percent died on campaign. At least 120 men are known to have been executed by Loyalist troops, and 544 others mutinied and were transported to Jamaica. Desertion was common, and personnel turnover was very rapid. Often a good soldier—whether local or foreign—was simply one who did not run away in the night.[56]

In some conflicts foreign volunteers held negative views about the fighting capabilities of their hosts while in others the hosts complained that the foreigners who had come to help them were more trouble than they were worth. Henry Noel Brailsford's impressions of the Greeks he served with in the war against the Ottomans in 1897 exemplify the former while Boer appraisals of their foreign volunteers in 1899–1900 typify the latter. The Scotsman Brailsford served on the front between Larissa and Pharsala, today in central Greece, for a few weeks in spring 1897. He watched a senior Greek commander and other officers abandon their troops and was obliged, together with other foreign volunteers, to fire upon panicky Greek soldiers trying to displace women from a train carrying civilians and the wounded to safety. In another engagement he saw local soldiers crouched behind rocks without firing a shot. Witnessing the collapse of the Greek effort, he soon became disappointed with their cause and returned to Britain.[57]

Two and a half years later, the young state attorney of the South African Republic (and future leader of South Africa) Jan Christiaan Smuts drew up a memorandum in preparation for the war he foresaw with Britain. In it he implored the Boer leadership to invite military experts from Europe

> to give advice which could be of the greatest value to our armies. This is specially necessary because our people have until now been used to fighting in small numbers by means of commandos, and in the struggle that we are approaching, it will be essential to fight in thousands.[58]

However, once the war began, Smut's vision of learning from foreign experts did not materialize. The commandant-general of the Transvaal, Piet Joubert, complained to his government of the troubles caused by the foreign volunteers. Many volunteers did not speak the Boers' language and were unsuited for their style of warfare. The costs of maintaining them were high. Some of them did not have proper training and were killed without furthering the Boers' cause. The rank and file too were mostly contemptuous toward the foreign volunteers. Those who could not ride a horse or shoot were ridiculed. Foreigners, especially Dutch volunteers, who joined commando units were either ignored or insulted. Only those who displayed physical fitness or martial aptitude that was equal or superior to that of their hosts managed to earn the locals' respect.[59]

Historians have tended to be dismissive of the military performance of the Arab Liberation Army (ALA), established by the Arab League in late 1947 and dispatched to fight against the Jews in Palestine in early 1948. Composed of volunteers from Syria, Iraq, Lebanon, Transjordan, and Palestine itself, as well as some exiles from Yugoslavia, it was led by seconded or retired officers, mainly from the Syrian and Iraqi armies. The ALA established its training camp at Qatana near Damascus and was, nominally, under the supervision of the Arab League Military Committee. In total, the ALA included approximately 6,000 men, some of whom were posted to Jaffa and Jerusalem while others joined small Palestinian bands elsewhere. The largest contingent, some 3,000 to 4,000 strong, was led by the veteran soldier of Arab nationalism, Fawzi al-Qawuqji. In February 1948 the Arab League put Qawuqji in charge of the war effort in northern Palestine. Qawuqji's troops were better equipped than the local Palestinian bands. For instance, they had a few artillery pieces and were able to bombard the besieged Jews of western Jerusalem. However, the ALA failed in its attempts to conquer Jewish settlements, most notably Tirat Zvi and Mishmar Haemek.[60]

In his memoirs, Qawuqji explained the failures of the ALA first by constant British interventions that saved the Jews every time they were about to be defeated (before British forces completed their withdrawal from Palestine in mid-May they intervened in the fighting on several occasions). Later he pinned the blame on "the scandalous lack of arms, ammunition, rations, clothing, health services and means of communication," which the Arab League Military Committee in Damascus failed to provide. He also complained of the poor training his men had received, saying that some of them could not even load a rifle properly.[61] ALA units, as Jewish intelligence reports indicated, suffered from acute discipline problems, with soldiers complaining that salaries were very low and that the food they received was insufficient.[62] At the same time, some of the foreign volunteers were disappointed by the lack of commitment of local Palestinian Arabs to the fight against the Jews. In turn, some of the Palestinian Arab population resented the presence of these foreign contingents. Hence, relations between ALA forces and their "hosts" were strained in Jaffa, in the Jerusalem area, and elsewhere.[63] Qawuqji later lamented "the neglect, ambition, shortsightedness and bad training of the senior officials who were dealing with the Palestine problem," but, as a fighting force, the ALA suffered from all the shortcomings that normally render a military force ineffective.[64]

Estimates regarding the military performance of foreign volunteers in Afghanistan in the late 1980s and in Croatia in the early 1990s are less clear cut than the general consensus regarding the ALA in 1948. Abdullah Anas, one of the founders of the Services Bureau that took care of in-coming volunteers in Pakistan, estimated that only a few hundred foreigners were actively engaged in Afghanistan at any given time. The rest, up to 90 percent of the total, remained in Pakistan and worked with Afghan refugees as teachers, doctors, cooks, and so on. In his eyes, some of the non-combatant volunteers were no more than tourists:

> Most of the [foreign] mujahideen were Saudi. They had money and sometimes they used good hotels, good restaurants. Some of the Saudis used to come and go as a trip, as a holiday. Most of them are coming for the holidays, but Islamic holidays. . . . [They] would go spend a few days or weeks with the Afghan refugees; sharing their suffering, bringing some clothes, some money.[65]

Some of the Afghan Arabs, therefore, earned a reputation for being "Gucci Jihadists."[66] However, there is also evidence that some of the foreign volunteers were highly committed to the armed struggle, fought shoulder to shoulder with their Afghan hosts, and sustained heavy casualties. This was the case in the Battle of Jaji in eastern Afghanistan in the spring of 1987. It took place near the so-called al-Masadah (the Lion's Den) camp, which Osama bin Laden had deliberately constructed in close proximity to a Soviet base. There are different accounts of what exactly happened in this battle and of the size of the foreign contingent that took part in it. There is little dispute, however, about its intensity. According to one version, one-fifth of the foreigners who took part in the battle were killed while trying to attack Soviet positions.[67] The former Afghan Arab Mustafa Hamid described the face-to-face fighting between the Arab "foot soldiers" and Soviet Special Forces in Jaji as a series of "dazzling victories, the likes of which had not previously occurred and were not repeated with the Arabs in the rest of the war."[68]

Having achieved notable success in guerrilla warfare, the Afghan muja-hideen and their foreign volunteers fared badly when it came to pitched battles. In March 1989, shortly after the Soviet withdrawal, a local force, which included a foreign-volunteer contingent led by bin Laden, attempted to cap-ture Jalalabad from the pro-Russian Afghan government. The American pho-tographer Steve McCurry, who covered the battle, observed that "the Arabs . . . were clearly willing to be on the front lines, willing to sacrifice." The attack, however, was comprehensively defeated, and bin Laden's force lost approximately 100 men, more than in the previous years of the conflict combined.[69]

The foreigners who joined Croatian units during the 1991–1995 war varied greatly from each other in terms of their military contribution. Some of them were troublemakers and criminals, as we shall see in Chapter 7. Others were described by veterans as "café warriors" because they steered clear of the frontline while basking in the prestige of being a foreign volunteer in uni-form.[70] On the other hand, some volunteers had extensive military back-ground prior to their arrival in Croatia. Such volunteers took part in heavy fighting against the Yugoslav National Army and various Serb units, as the story of Antaine Mac Coscair demonstrates.

Born in Manchester, Britain, to a Catholic family of Irish descent, Mac Coscair attended a naval school before enlisting as a junior leader in the Royal Artillery at the age of 16. He went on to serve in an antiaircraft regiment in West Germany, leaving the British armed forces at the age of 20. News cov-erage of the war in Croatia made him feel "a strong sense of injustice," so he left his security job and made his own way to Zagreb, arriving in August 1991. He then spent the next several months fighting in hastily organized units of the Croatian National Guard (ZNG), the Croatian Defense Forces (HOS), and eventually the Croatian army (HV). When the Croats launched their of-fensive to liberate Western Slavonia in late October, Mac Coscair joined a reconnaissance unit that carried out commando raids against the Yugo-slav National Army base at Rajčići and suffered heavy casualties from the enemy's artillery and aerial bombardment. In mid-November Mac Coscair and a fellow foreign volunteer transferred themselves to a highly profes-sional twelve-man special unit that had been set up in October by John "Olly" Scarrott, 44, a former member of the Royal Marine Commandos, and Tom Crowley, a 43-year-old who was believed to be ex-Rhodesian Special

Air Service. Other team members included two more Royal Marine commandos, four former French Foreign Legion paratroopers, three who had served with the regular British army and one former soldier from the Royal Danish Army. The average age of group members was 31. The unit's activities included reconnaissance, mortaring, and sniping at enemy positions. Ambushed on 18 November 1991, they managed to extricate themselves from a difficult situation, carrying away two wounded members.[71] This particular unit was better trained than the majority of the local troops. However, because of the small number of highly professional foreigners, their impact was restricted to the level of tactical engagements.

The evidence on the tactical effectiveness of foreign contingents in newly established military formations such as the one led by Bolívar in the early nineteenth century, the Philhellenic Legion in Greece in 1897, the Boer forces in 1899–1902, the ALA in 1948, the Afghan Arab cohort in the 1980s, or Croatian units in the early 1990s is inconclusive. None of these foreign contingents could claim to have significantly altered the outcome of the conflict they participated in. On the basis of these examples, it seems highly unlikely that foreign war volunteers increase the intensity, duration, and intractability of conflicts, as UN Security Council Resolution 2178 suggests. But yet there were other ways—beyond the tactical level of regular, irregular, and hybrid battlefield formations—in which volunteers from abroad managed to make an impact on the conflicts they took part in.

A Qualitative Edge: Training and Technical Expertise

As Millett and Murray observe, effective leadership in times of war entails not only access to manpower in sufficient quantities but also of the right quality. One realm where foreign volunteers with previous military experience could act as force multipliers is in training new recruits. It is therefore not surprising that foreigners have often been used in training capacities in various conflicts. When the former US Army colonel "Mickey" Marcus first arrived in Palestine in early 1948, his initial task was to advise Jewish forces on military command, control, and training. Many of his suggestions dealt with logistics, an area that the pre-state Jewish paramilitary force, the Haganah, undervalued. "An army isn't just about fighting," he would tell his hosts. "A soldier's shoes and food are as important as his ammunition."[72] Willy

Van Noort, a retired colonel from the Dutch army, trained new recruits in Croatia during the war of 1991–1995. In early 1992 he and a few other foreigners taught Croatian troops in the Sisak area how to lay an ambush, use different kinds of weapons, and improve their fitness through ever-increasing long marches. General Slobodan Praljak, the Croatian assistant secretary of defense during the war, spoke highly of Van Noort in a testimony he gave in 2009.[73] Samir al-Suwaylim (better known as Ibn Khattab or Emir Khattab), a Saudi-born veteran of the war in Afghanistan, was asked by the leadership of the Chechen insurgents to set up training camps for new recruits in 1996. Many of the men who passed through these camps went on to fight against Russian forces in the Second Chechen War from 1999 onward.[74]

Experienced foreign volunteers could act not only as conveyors of knowledge through training but also as recruiters. It has often been pointed out that veterans of the Afghan conflict in the 1980s went on to recruit new jihadists to fight in Bosnia and in other hot spots.[75] Gaston Besson, a French volunteer who took part in the fighting in Croatia and Bosnia during the 1990s, was reportedly overseeing the enlistment of foreigners to the Azov Battalion in Ukraine in 2014.[76]

In a small number of cases, the skills of foreign volunteers made a more discernible qualitative contribution to the conflicts they fought in, as the following examples from Poland, Israel, and Iraq illustrate. Captain Merian C. Cooper was a young American Air Service officer from Jacksonville, Florida, who had flown in France during the First World War before being held as a prisoner of war by the Germans. In the winter of 1918–1919 he served in the American Relief Administration in Poland. Stirred by clashes with Ukrainian forces in eastern Poland, Cooper volunteered his services and offered Polish authorities to travel to Paris and recruit a squadron of trained pilots. He argued that, while the newly independent Poland could train its own pilots, he would be able to enlist "men who could bridge the gap of the inevitably slow and costly evolution of fighting and defense units."[77] Cooper kept his word. Over the summer of 1919 he recruited First World War veterans who had served in France and Britain.

Before going into service in October of that year, the foreign recruits met with Marshal Jozef Pilsudski, who was keen neither about airpower nor about foreigners. "Poland is able to fight her own battles," he told them. Once they managed to convince him that they were committed to the cause and were willing to receive equal pay to Polish officers of corresponding rank, Pilsudski relented: "You will, I fear, have great difficulties with our language; our aviation equipment is at present almost nil; you will meet with difficult obstacles at every turn."[78]

The most notable contribution of these foreign pilots was in providing aerial reconnaissance and strafing the so-called Red Cavalry of the Soviet commander Semyon Budyonny, which was advancing westward into Poland in the summer of 1920. One of the pilots, Kenneth Malcolm Murray, who wrote a book about the history of the squadron, conceded that "just how many casualties the pilots had caused among the Cossacks there was absolutely no way of ascertaining." Yet he believed that, between the combined action of the ground and air forces, the first great impetus of Budyonny's advance "had unquestionably been seriously checked."[79] In support of his assertion, Murray reproduced an intercepted radio message sent by Soviet commanders: "In the last battle near Lwow the troops of the Cavalry Army suffered great losses due to the airplanes of the enemy."[80] However, this statement should be qualified. Budyonny needed to justify his setbacks, and blaming air power, which his forces lacked and could not be expected to compete with, would have been a convenient scapegoat.[81] Nevertheless, it is clear that the newly established Polish armed forces were able to gain a qualitative edge over their enemy by enlisting experienced foreign personnel.

Similarly, the most substantial qualitative contribution that the newly established State of Israel was able to derive from its foreign volunteers was in the creation of an air force. Starting in the autumn of 1947, Zionist emissaries from Palestine worked alongside highly committed diaspora Jews to recruit Second World War veterans. The Israeli defense ministry representative in London, Rafi Thon, estimated at the end of the war that up to 70 percent of the volunteers from Britain had had previous military experience. Pilots with plenty of flying hours were especially in demand. When necessary, the recruiters agreed to pay handsome salaries to those pilots who were otherwise reluctant to commit.[82] Thus, 171 of the 193 pilots who flew combat missions for the Israelis in the 1948 war were foreign. Other volunteers, primarily from English-speaking countries, served as transport aircraft pilots, air crew members, and ground technicians.

Foreign volunteers in the air force fulfilled three central tasks that greatly assisted the Israeli war effort. Their most important contribution was in flying war matériel into the country. Before the end of the British mandate in Palestine in mid-May 1948, aircrafts smuggled firearms and ammunition into territories held by the Jews, mainly from Czechoslovakia, and had to land at night at improvised or barely-lit airfields to evade the British military. Procuring weapons remained a problem even after the British withdrawal and the proclamation of Israeli independence on 15 May. The United States had already placed an arms embargo on Palestine in December 1947. UN Security Council Resolution 50, adopted on 29 May 1948, called upon member states to refrain from exporting war matériel into Palestine and other Middle Eastern countries that had joined the war. Britain, the traditional weapons supplier for Egypt, Jordan, and Iraq, abided by this arms embargo, as did France, which had previously supplied Syria and Lebanon. The Czechoslovak government however, with the approval of the Soviet Union, agreed to continue supplying weapons for the Israelis. The Israeli air force was therefore able to fly in heavier weaponry such as artillery pieces, armored vehicles, and disassembled aircraft while the Arab states struggled to replenish their stockpiles.[83]

A second major contribution of foreign personnel, especially those who were veterans of the Second World War, was in helping to train new recruits to become pilots and to fill all the other positions that a modern air force required.[84] A third form of contribution was flying supplies to isolated Jewish settlements that could not be accessed otherwise and bombing enemy convoys, positions, and cities. It should be noted that battles in the war of 1948 were not decided in the air. The Israelis, on the one hand, and the Arab states combined, on the other, had no more than a few dozen serviceable aircraft each. That the Israelis were able to achieve complete air superiority over the Egyptian air force by October 1948 was not the result of an advantage in numbers of aircraft or their particular quality. While the Egyptians suffered from a lack of staff and spare parts, the IDF's personnel were able to maintain aircraft for use around the clock.[85] The key was in the interaction between skill and technology, a central facet in mechanized warfare.[86]

Foreigners who joined militant Sunni organizations in Iraq in the years following the US-led invasion of 2003 had a very different, though highly noticeable, impact on the local insurgency. Although the vast majority of insurgents were local, the leadership of al-Qaeda in Iraq was overwhelmingly foreign.[87] According to the political scientist Mohammed Hafez, foreign ji-hadists imported the practice of suicide bombings into the Iraqi conflict, had specialized skills in manufacturing explosives, and convinced local recruits to adopt this tactic. In addition to the small number of experienced veterans— such as al-Zarqawi—who operated as commanders and planners, foreign jihadists also supplied the bulk of the foot soldiers who carried out suicide attacks.[88] A cache of insurgent records that was found in Sinjar in 2007 indi-cated that more than half of the foreign fighters who entered Iraq through Syria were designated to become or saw themselves as potential suicide bombers.[89] In a way, suicide bombers replaced the cannon fodder of old. They did not require lengthy military training or specialized skills, only the com-mitment and resolve to kill themselves in the midst of their target.[90] Yet, with Iraqi security forces and civilians constituting the prime targets, few would deny that suicide attacks were a major precipitating factor in the sectarian spiral of violence between the Sunni and Shia of Iraq. In May 2005, for in-stance, more than 1,200 people were either killed or injured by suicide attacks alone.[91] As US Air Force general Don Alston stated in the summer of that year, "the foreign fighters are the ones most often behind the wheel of suicide car bombs, or most often behind any suicide situation."[92]

Of course, there are differences between these examples. In the state-on-state wars fought by the Poles in 1920 and the Israelis in 1948, the national leadership was able to benefit from military experience and expertise that had been acquired elsewhere and that complemented their military needs. In the Iraqi example, the foreigners invited themselves to take part in the local insurgency with little regard for what the local population wanted.[93] How-ever, as all three examples illustrate, when foreigners brought with them skills, operational concepts, and organizational structures that were suc-cessfully integrated into their hosts' fighting machine, they could have a discernible impact on conflicts, even though they constituted a small mi-nority within the local forces. Such cases remain rare. While states and political entities, most notably in Africa, have been able to acquire mili-tary or technical expertise to plug holes in their armed forces through the use of mercenaries and private contractors, few hosts have been in a posi-tion to attract foreign volunteers with precisely the skills and combat ex-perience they needed.

Propaganda and Morale

The historical record suggests that the most consistent contribution of foreign volunteers has been political rather than military. Examples of how the presence of foreign volunteers could be used to boost domestic morale or harnessed for propaganda targeted at international audiences are plentiful. In 1875–1876 Prince Nikola of Montenegro hoped that by hosting Garibaldian volunteers he could gain international support for his struggle against the Ottoman Empire. Some of the Italian volunteers who went to the Balkans certainly helped in raising attention abroad by sending reports that were published in newspapers back home. Contemporaries and historians alike tended to dismiss the military value of the Garibaldian volunteers in Herzegovina and elsewhere in the Balkans, focusing in their appraisals on the foreigners' propagandistic role.[94] Future generations of men who followed in the Red Shirt tradition found themselves appropriated for political purposes to an even greater extent. In the spring of 1915 some of the demobilized Garibaldian volunteers who had joined the French Foreign Legion and fought against Germany returned to Italy and were soon seized upon by politicians and publicists that hoped to bring neutral Italy into the First World War. Interventionists such as the future dictator Benito Mussolini and the poet Gabriele D'Annunzio paraded veterans of the fighting in northeastern France in the public manifestations they organized.[95]

Flora Sandes, the British nurse-turned-soldier who joined Serbian forces during the First World War, helped to raise the locals' morale when the situation seemed bleak and also worked to raise awareness of their cause internationally. When she decided to take up arms and joined the ranks of the Serbian army, she was told by her commander that, "if I stuck to them it would encourage them, and strengthen their belief that in the end England would help them."[96] During her wartime service she traveled twice to London to collect funds for Serbia and also published an early account of her experiences, *An English Woman-Sergeant in the Serbian Army*, in 1916.

In a similar vein, Salaria Kea, the nurse from the United States who was injured during her service in the Spanish Civil War, contributed to fundraising activities upon her return to New York. She traveled through the country on behalf of the Medical Bureau of the American Friends of Spanish Democracy, urging people to aid Republican Spain with money, medical aid, and food. Her story was celebrated in a pamphlet issued by the Negro Committee to Aid Spain.[97] Other injured foreign volunteers who served in Spain, such as the German writer and political commissar Gustav Regler, participated in similar fund-raising tours in the United States.[98] In some instances foreign volunteers also assisted in raising local Spanish morale. According to various accounts, the first arrival of the International Brigades in Madrid in November 1936 galvanized the Republican defenders of the capital, who were under attack from General Franco's forces.[99] In her famous farewell speech to the International Brigades, delivered in Barcelona on 1 November 1938, the communist leader Dolores Ibárruri, also known as "La Pasionaria," recalled, "the hardest days of the war, when the capital of the Spanish Republic was threatened, it was you, gallant comrades of the International Brigades, who helped save the city with your fighting enthusiasm, your heroism and your spirit of sacrifice."[100]

Foreign volunteers were also used for propagandistic purposes during the Second World War. The British ambassador in Washington, Lord Lothian, as well as officials in the Ministry of Information sought to use individual American pilots and, later, the RAF Eagle Squadrons to boost support in the United States for Britain in 1940–1941. Because the pilots who were cast for leading roles, both on- and off-screen, were still in active service, the plans of the propagandists were often thwarted by the rude intervention of the war. William Fiske III, a 29-year-old originally from New York, and half a dozen other pilots from the United States were killed before their propagandistic contribution materialized. The Ministry of Information was nonetheless involved in the production of the films *A Yank in the RAF* and *Eagle Squadron,* compelling the RAF to cooperate with the filmmakers.[101]

Contemporary readers will undoubtedly be aware of the propagandistic importance of jihadi foreign fighters. As the political scientist Barak Mendelsohn observed in early 2011, "Western volunteers, many of them more technologically savvy and less capable as fighters than their local brothers-in-arms, are now being channeled to roles in which their qualifications could best be utilized."[102] Foreign volunteers have helped to attract international attention to the conflicts in Afghanistan in the 1980s, Bosnia in the 1990s, Iraq and Somalia in the 2000s, and, to an even a greater extent, the Syrian Civil War in the 2010s. They released statements and videos, translated messages into English and other European languages, appealed to Western Muslims, threatened Western governments, and operated social media websites. One need only think of the international repercussions of the filmed executions of Western hostages carried out by "Jihadi John" (Mohammed Emwazi) to realize the propagandistic impact of foreigners who have joined the Islamic State.

The targeted killing of Junaid Hussain in al-Raqqah in the summer of 2015 provides a clear testament to the importance attached in the West to the on-line activities of foreign fighters in the Syrian conflict. According to the *New York Times,* Hussain, a 21-year-old from Birmingham in the United Kingdom, was the leader of a small group of English-speaking Islamic State propagandists, hackers, and recruiters nicknamed "the Legion." Hussain and his associates, many of whom were killed by American and allied drone strikes, sought to recruit and inspire Muslims in the West to carry out terrorist attacks.[103]

The foreign volunteers who joined Kurdish units to fight against the Islamic State have also had a noticeable propagandistic value. When the actor Michael Enright arrived in Syria he did not speak the local language and had no prior military training. Militarily, he was of very little use to his hosts. Nonetheless, the interviews given by Enright and other foreign volunteers helped to bring international attention to the struggle of the Syrian Kurds. As former Pentagon analyst Phillip Lohaus observes, "Putting a Western face on a foreign problem helps make the conflict more relatable to people in the West."[104]

* * *

The evidence examined in this chapter suggests that the recruitment of foreign volunteers does not pave their hosts' way toward victory. Foreigners have invariably been too few to win wars on their own. No generalizations can be made in terms of their commitment to the fight. In fact, there are examples of volunteers being more committed, as committed, and less committed than their hosts.

The evidence also shows that, historically, it has been easier to integrate foreign volunteers into newly established military formations where organizational structures tend to be more flexible and fluid. However, over time, such formations often underwent a dual process of nationalization and centralization, especially in cases of protracted conflicts. For instance, when the Spanish Civil War began, an array of militias appeared on the Republican side and some of these incorporated foreign volunteers. Between the autumn of 1936 and spring of 1937, the republican government worked hard to combine its forces into the Popular Army (Ejército Popular). The International Brigades—initially an autonomous force—were eventually formally integrated into the Popular Army in September 1937.[105] In late 1991, following a series of clashes and standoffs between forces loyal to the Croatian government and HOS units (where many of the foreign volunteers served), the latter were forcefully amalgamated into the newly established HV.[106] When the war spread to neighboring Bosnia-Herzegovina in 1992, a number of militias were established in the areas loyal to the Sarajevo government. These were later replaced by more organized units of the Bosnian army, the composition and ethos of which became increasingly Bosnian-Muslim.[107] In October 2015 the Rada, Ukraine's house of representatives, passed a law allowing foreigners to enlist in the Ukrainian army. Until that point foreigners had served in semiofficial outfits such as the Azov Battalion. The new law paved the way for greater centralization and government control.[108] These processes of centralization and nationalization inevitably diluted the importance of foreign volunteers.

In state-on-state wars and in wars of secession, there is little evidence to support the assumption that foreign volunteers have increased the intensity, duration, and intractability of conflicts. In conflicts characterized by guerrilla or irregular warfare, because of their fluid structures and relatively small size of insurgent forces, foreign volunteers have a greater potential to make an impact. In Iraq it could be argued that the practice of suicide bombings, introduced by foreign Sunni commanders, exacerbated the conflict and gave it a distinctly sectarian character. In Chechnya foreign fighters have contributed to the fragmentation of the anti-Russian movement between nationalist and ultrareligious groups. This fragmentation weakened rather than strengthened the insurgency, though it may have also prolonged the conflict by making a mediated solution more difficult.[109]

In what Laia Balcells and Stathis Kalyvas have called "conventional civil wars," characterized by pitched battles, urban sieges, and clearly defined frontlines, the picture is open to interpretation.[110] It has been argued that the International Brigades prolonged the war in Spain by saving Madrid in late 1936, but recent studies have emphasized other factors that brought about the Republican victory in that battle.[111] More work needs to be done to assess the extent to which foreign fighters in the civil war in Syria have contributed to its intensity and length.[112] It is clear that some foreigners have played prominent roles in the conflict, such as Georgian-born Tarkhan Batirashvili (better known as Omar al-Shishani), who is believed to have commanded Islamic State forces in northern Syria. But, at present, the fog of war (and information warfare) is far too thick to properly discern the discrete military significance of the foreigners in Syria more generally.

Nevertheless, on one thing different types of warfare converge. Only in conflicts where the hosts were able to attract foreign volunteers with skills and combat experience that their enemies did not have could they derive benefits from their guests which exceeded the tactical level. In most cases, the main contribution that foreign volunteers have made was in the realm of propaganda and their impact has been more political than military. But while transnational recruits could help to raise local morale and amplify international media coverage, thus benefiting their hosts, they could also have very adverse effects. As the next chapter illustrates, individual foreign volunteers have had the propensity to cause a lot of harm both during the wars they fought in and in their aftermath.

* 7 *

The Dark Side

Troublemakers, Soldiers of Misfortune, and Terrorists

At the very end of *Anna Karenina*, Leo Tolstoy takes his readers to a Moscow train station in July 1876, where a large number of volunteers are about to depart for the war in Serbia, to fight against the Turks. Some time has passed since Anna's suicide, and a repentant Prince Vronsky not only volunteers his services but also takes with him a whole squadron at his own expense. Many flock to the station to bid the volunteers farewell. Money is collected for the cause and a speech is given to celebrate the volunteers' service to "the Faith, humanity, and our brothers!"[1] Yet when one of the characters, Katavasov, boards the train and converses with the volunteers, an unflattering picture begins to emerge. The group includes a failed merchant, bad soldiers, and unsuccessful retired officers, many of them drunk. One of the men struggles to give convincing reasons for his decision to volunteer. Katavasov is fearful of speaking openly against the volunteers because their cause was so popular, reflecting the great Russian author's own lack of enthusiasm for the Pan-Slav movement. He viewed the volunteers of 1876 as men who, like Vronsky, were running away from something.[2] As Tolstoy subtly alluded, there was often a gap between the image of the heroic, selfless volunteers and the real-life individuals who stood behind this image.

Tolstoy is not alone. While some foreign volunteers have been wrapped in a romantic aura, many others have been described as mercenaries and thugs, especially by their enemies.[3] Even if we set aside biased and partisan appraisals, we will find that in conflicts where foreigners have taken up arms, idealists for whom the moral pull was strong were accompanied by opportunists and lawless adventurers. For some of these, their search for meaning found expression in the desire to go to war for war's sake. Consequently, wanton violence, criminality, excessive drinking, and so on have been an almost unavoidable corollary of voluntary transnational military service. As we will see, the behavior of some volunteers has produced bitter disappointment among both their hosts and their peers. Another negative aspect of the foreign-volunteer phenomenon relates not so much to the volunteers themselves as to their hosts. In fact, in some cases well-intentioned foreigners who arrived in a war zone with a desire to help soon encountered a reality that was very far removed from what they had expected. Ultimately, this chapter illustrates how participation in a foreign conflict could generate both disillusionment and what we would now call "radicalization," though not in equal measures.

Troublemakers

"I always wanted to kill legally," a British man called Dave told the journalists making the documentary film *Inside Story: Dogs of War* in eastern Croatia in early 1992. He adored the idea of having no compassion toward his victims. The camera later shows the helmet of this former nightclub bouncer adorned with the words "Yorkshire Ripper."[4] One wonders whether this is the sort of defender the Croatian cause—or any other cause, for that matter—would wish upon itself. Yet this display of warped masculinity and lust for violence is a grim reminder of the sense of empowerment and altered perception of what is permissible that service in a foreign war can bestow.

Wartime empowerment could unlock violent desires even among ideologically committed volunteers. The Communist activist Joseph Dallet, for instance, was dedicated to the cause of the Spanish Republic ever since the outbreak of the civil war in 1936. Before leaving the United States, he was involved in efforts to raise money and send food and medical supplies to Spain. After "fighting within the [Communist] Party to get permission to go there," and having twice been refused, he finally set out in March 1937.[5] He was arrested en route to Spain and subsequently spent twenty days in a

French prison. After his release, he managed to smuggle across the border in April of that year. Once in Spain, he was able to express the satisfaction he derived from military life in a letter to his wife, Kitty:

> Man what a feeling of power you have when entrenched behind a heavy machine gun! . . . You know how I always enjoyed gangster movies for the mere sound of the machine guns. Then you can imagine my joy at finally being on the business end of one.[6]

Dallet, whose commitment to the cause cannot be questioned, was killed on the Ebro front while trying to storm an enemy position. Any admiration he may have had for gangsters remained a fantasy. Foreign volunteers responded to or made use of the wartime conditions they encountered in different ways, some of them more similar to Dallet, others more similar to Dave. The following examples explore the more negative side of the spectrum. These include the propensity of some volunteers toward dueling or brawling, lying about their pasts, stealing or committing other crimes, and, occasionally, betraying their peers.

Accounts depicting the philhellenic volunteers, who fought against the Ottoman Empire in the 1820s—whether written by veterans or by historians—emphasize how honor or the lack thereof framed the conduct of many of these foreigners. Volunteers from France, the German states, the Italian peninsula, and other parts of Europe began to arrive in Greek ports such as Navarino (Pilos), Kalamata, and Missolonghi in 1821. Soon after their arrival, duels over points of honor broke out, abetted by heavy drinking. Hence, a few foreigners were injured almost as soon as they landed and remained unfit for further service. In February 1822, a pigsty in Navarino was taken over as a place of punishment into which the drunken and unruly could be thrown. A number of volunteers lied about and exaggerated their previous military backgrounds. One Frenchman, known as Mari, arrived in Corinth and passed himself off as a former officer in Napoleon's guard. Before long he disappeared, along with a handful of other volunteers, and joined the Turkish army. Another Frenchman, Captain Jourdain, promised Greek leaders in the Argolic Gulf area that he was able to set fire to a Turkish position. However, once the Turkish fleet appeared and the alarm was sounded, Jourdain jumped into a boat and disappeared.[7]

In May 1822 the Battalion of Philhellenes, composed of two companies of about fifty men each, was formed in Corinth. The practice of dueling did not disappear, however. En route to Peta, where the battalion would take part in a disastrous battle, the French and German companies quarreled over a killed sheep, which the former refused to share with the latter. The matter was settled in a duel in which the French champion prevailed over his German counterpart. Even after this event, tension between the two national contingents remained high and further duels and scuffles followed.[8] In principle, the philhellenes had gone to Greece to fight for liberty and against Ottoman oppression. However, as the examples above illustrate, some of these foreigners ended up fleeing or fighting among themselves. Subsequent conflicts provide further, though not necessarily similar, examples of volunteers whose motivations and conduct were very questionable.

In February 1940, during the war between the Soviet Union and Finland, the Finnish Foreign Ministry received a number of warnings not to recruit a Spaniard identified initially as Colonel Alfonso Reyes. Arthur Vivian Burbury, a British Foreign Office official who had previously been posted to Finland, was first to sound the alarm. He sent the Finnish minister in London information about Reyes, which had reached him from various sources, so that the Finns could "avoid the possibility of any dangerous deception." The information was patchy, contradictory, and not very credible. Said to be a bigamist and a scoundrel, marrying rich women for their money, Reyes had allegedly served time in prisons in Spain, France, the United States, and Japan. On the one hand, Reyes was reported to have served in the Spanish Foreign Legion and to have been a personal friend of Franco. On the other hand, he was also described as a Communist who fought for the Spanish Republic during the civil war.[9] Later reports that reached the Finnish Foreign Ministry from Stockholm described Reyes as the former chief of the Spanish Republican air force.[10] His supposed previous associations with Soviet pilots raised the concern that Reyes, also described in some reports as "Beres" or "Perez," might be a spy. The available sources suggest that Reyes—if that was indeed his name—never saw active duty with Finnish forces. Having passed through Sweden in early 1940, he was believed to be in Denmark in late April.[11] This example nonetheless illustrates how the prospect of traveling to join a conflict abroad could have a strong appeal for troublemaking adventurers.

As the example of Dave, the former bouncer with the "Yorkshire Ripper" helmet, illustrates, the conflict in the former Yugoslavia in the 1990s attracted a number of unsavory characters from abroad. In early 1992 Dave was arrested for stabbing a German journalist in a barroom brawl. Dave's commander was Eduardo Rózsa-Flores, a polyglot Hungarian-Bolivian who went to Croatia as a journalist but soon decided to take up arms. Thanks to his good relations with local Croat commanders, Rózsa-Flores was given command of the First International Company on the Osijek front. The British journalist John Sweeney later accused Rózsa-Flores and one of his soldiers, Stephen "Frenchie" Hancock, of being involved in—if not responsible for—the deaths of the Swiss journalist-turned-volunteer Christian Würtenberg and the British photographer Paul Jenks (both of whom were killed in January 1992).[12] Rózsa-Flores and Hancock both denied these allegations.

The picture of Rózsa-Flores's international unit that emerges from two documentaries that were filmed during the war—*Inside Story: Dogs of War* (1992) and *Travels with My Camera: Dying for the Truth* (1994)—is a very negative one.[13] Hancock, for instance, openly admits to blowing up houses that belonged to Serbs in Osijek to serve as a warning for others not to collaborate with the enemy. Speaking about his motivations for fighting in Croatia, Kit, who formerly served in the British army, said: "We don't make good civilians. . . . We are forever breaking laws." Another British member of the unit, Carl Finch, also known as Karl Penta, had served as a mercenary in Suriname and in other hotspots during the 1980s.[14] Halfway through the filming of *Inside Story: Dogs of War*, Penta and two other of the unit's instructors were locked up by military police after allegedly having a drunken confrontation with Rózsa-Flores. Yet another British volunteer later claimed that he and a companion tried and failed to shoot Rózsa-Flores in the back during a skirmish with the Serbs near Osijek.[15] There were also a few troublemakers among the foreign volunteers fighting on the Croatian side further south, in the Vinkovci area, during the same period. One of them was arrested for murdering a local taxi driver in January 1992.[16]

In March 1993 the American Rob Krott returned for his "second tour" in the Balkans.[17] His destination was the town of Tomislavgrad in western Bosnia, where he joined the King Tomislav Brigade, part of the Bosnian-Croat armed force, the Hrvatsko vijeće obrane (HVO). Approximately twenty foreigners served in this predominantly local unit. Save for a small number of committed and professional soldiers, the foreign volunteers he encountered there were, in his opinion, a ragtag of misfits and adventurers. Many of them routinely lied about their military background, claiming to have served in elite units in their national armies or in the French Foreign Legion. In his retrospective account, Krott says he had felt as though he was "surrounded by criminals and amateurs" as well as "thieves, deserters and troublemakers."[18] He describes a couple of the German volunteers in his unit as skinhead neo-Nazis, one of whom had previously spent time in a German prison as a result of a race-related crime. Another had bombed a Communist bookstore in Germany. Still wanted by the police, the latter traveled using someone else's passport.[19] Krott nonetheless had more respect for these Germans than he had for "James," an American volunteer who claimed to be a former Marine: "Of only moderate intelligence, and somewhat emotionally disturbed, he fell into the 'I can do anything I want here' mode." This behavior was typical, he said, of those who "think they can be a law unto themselves, and gravitate to places like Bosnia just for that purpose."[20]

Pay could not have been a motivation for joining the King Tomislav Brigade since salaries were irregular and extremely low by Western standards. Receiving the equivalent of $2 to $3 a day, the foreigners referred to their wages as "beer and cigarette money."[21] To compensate themselves, some volunteers took part in looting and other transgressions. With a cease-fire on the horizon and growing increasingly wary of the "crackpot criminal schemes" of some of the foreigners who served with him, Krott decided to leave.[22]

An even more disturbing image of the conduct of some of the foreigners in the King Tomislav Brigade emerges from the Finnish-language book *Mostarin Tien Liftarit* (Hitchhikers on the Road to Mostar). Published in 1997 under the pen name Luca Moconesi, the book's real author was Marco Casagrande, who went on to become a famous architect and artist. The book details conversations with neo-Nazi foreigners who proudly recount raping civilians, torturing prisoners, and carrying out other war crimes.[23] After the book's publication, Casagrande was questioned by Finnish authorities about any involvement he may have had with war crimes in Bosnia, but the author claimed that the book, or at least parts of it, were a work of fiction.[24] Roland Bartetzko, who briefly served alongside Casagrande in Bosnia in the summer of 1995, remembers him as a "smart and funny guy" who liked to talk a lot. Bartetzko's commander "didn't want to have him [Casagrande] in the unit as he was regarded as a 'tourist,' someone who spends his summer holidays in 'action' and after a couple of weeks goes back home." Casagrande talked his way into staying, only to abandon the unit after a short while.[25]

The presence of individuals with criminal pasts, or criminal tendencies, among foreign volunteers persists into the 2010s. Joe Robinson, a 22-year-old former soldier in the British army, had pleaded guilty to grievous bodily harm charges in August 2014 after fracturing a man's jaw in a brawl outside a bar. Given a suspended sentence, Robinson managed to slip out of Britain and traveled to the Middle East, where he fought alongside the Kurdish YPG against the Islamic State. After spending a number of months in Syria, he returned to the United Kingdom and was arrested upon arrival in November 2015.[26]

There are also numerous criminals among the foreigners who have joined the Islamic State during the civil war in Syria. A study compiled by a number of German security agencies, which included information on 677 individuals who left Germany for Syria or Iraq before 30 June 2015, found that 225 of them were suspected of or tried for criminal offenses prior to their departure. Assault and robbery, along with other property crimes and drug trafficking, were the most common offenses, while politically motivated crimes played no significant role.[27] If these individuals differ from their predecessors and contemporaries fighting in other conflicts, it is in the way they bring together a criminal past, voluntary military service abroad, and subsequent terrorist attacks against their home societies after returning from the war zone.

The story of Mehdi Nemmouche from Roubaix in northern France exemplifies this link in full. Nemmouche was a troublemaker even before he became a foreign fighter, spending five years in a prison in Lille on robbery charges. In 2012, three weeks after his release, he traveled to Syria. He spent about a year there, joining the ranks of the Islamic State where he served as a jailer. According to the testimony of the French journalist Nicolas Hénin, who was held hostage by the jihadist organization, Nemmouche and a small group of foreign volunteers would regularly torture prisoners captured by the Islamic State. Nemmouche also boasted to Hénin about other outrages he had committed, including the rape and murder of civilians. By May 2014 Nemmouche was back in Europe. He was arrested in France and charged with carrying out an attack on the Jewish museum in Brussels in which four people were killed.[28] With such a rogues' gallery, it is important to remember that not all of the foreign volunteers were villains. And sometimes their hosts were not exactly saints either.

Falling into Bad Company

Paradoxically, becoming a volunteer in a foreign war entails a certain degree of relinquishing control over one's fate. It is not only the risk of dying or getting injured. There is also an element of luck involved in the kind of company one falls into when reaching the host state or political entity. This is especially true in cases where foreign volunteers have joined insurgencies, civil wars, or wars of secession where a number of militias vie with each other and there is no institutionalized procedure for absorbing foreign recruits. The volunteers—because of their lack of familiarity with the host society—often find themselves in dubious units whose actions and politics they do not necessarily like.

Examples of such negative encounters emerged, for instance, from the reception of the philhellenic volunteers who traveled to southern Greece in the early 1820s. Those who arrived in Navarino early in the conflict were horrified to discover the remains of massacres carried out by Greeks against the town's Turks. In 1822 General Karl von Normann-Ehrenfels of Württemberg and his foreign contingent instituted a regular watch on the walls of Navarino to guard against a surprise enemy attack. The locals refused to take part, arguing that the Turks would not attack when it rained or under the cover of darkness. During the Battle of Peta in July 1822 a Greek commander deserted his section of the front, along with his men, enabling the Turks to flank the force headed by the Greek leader Alexandros Mavrokordatos, which included the Battalion of Philhellenes. Less than thirty of the foreigners survived, while sixty-seven were killed during the battle or in its immediate aftermath. The battalion was formally disbanded in late July and most of the survivors prepared to leave for home. Their commander, General Normann-Ehrenfels, who was wounded at Peta, remained in Missolonghi and died in November.[29]

For Abdullah Azzam and the other leaders of the volunteers who sought to take part in the Afghan struggle against the Soviet Union in the 1980s, one of the biggest hurdles was their hosts' tendency toward factionalism. Mustafa Hamid from Egypt was among the first foreigners to arrive. Recalling his visit in Pakistan and Afghanistan in 1981, he observed that "some of the Afghan leaders in Peshawar wanted to keep the Arabs away from Afghanistan and restrict their role to fundraising." This was because "if the Arabs went inside [Afghanistan] they could discover defects in the jihad situation, the most important of which was that the leaders in Peshawar were not the leaders in Afghanistan."[30] In fact, some of the early arrivals were drawn into disputes between rival Afghan factions. This not only led to the demoralization of volunteers but also increased the locals' mistrust toward their foreign helpers. To avoid, as far as possible, similar friction from emerging with the arrival of new cohorts of volunteers, Azzam and bin Laden introduced a cultural training program in the camps they established. The program sought to prepare new arrivals for service alongside the Afghan mujahideen. In the literature provided in the camps, foreigners were made aware of the rivalries between different Afghan groups and told not to believe what rival factions said about one another. Moreover, the volunteers were told not to spend too much time with only one faction at the front so as not to have themselves associated exclusively with that group.[31]

Factionalism was also a problem during the wars in the former Yugoslavia. Antaine Mac Coscair and a number of other foreigners were drawn into the feud between Franjo Tudjman's Croatian government and Dobroslav Paraga, leader of the ultranationalist Croatian Party of Rights along with its military wing, HOS. In November 1991 Paraga was arrested for allegedly plotting a military coup against the government. Subsequently, Mac Coscair and his teammates, though "unaware of the politics," found themselves defending the HOS headquarters in Zagreb for three days against an expected police raid.[32]

For some of the other foreign volunteers on the Croatian side, the main difficulty was not factionalism but the criminal behavior they encountered. Unlike Rob Krott, who left his unit because he did not trust the foreigners who served with him, the key problem for the British volunteer Simon Hutt were his local comrades. Hutt joined a HOS unit in northeastern Bosnia, near the Croatian border, in 1992. The unit largely consisted of Croat HOS members, though Hutt attached himself to a small group of foreigners which included Dutch, British, German, and American volunteers. Upon arrival, he was told that the local "gangsters and criminals are the minority [in the unit] but they're dangerous."[33] It was not long before he realized that some of the members of this nationalist paramilitary organization were avid neo-Nazis who would listen to recordings of Hitler. Alcohol and drugs were rife, as was black marketeering. "I'm not here for that kind of shit," Hutt soon told his new friends. Their response was: "That's why the foreigners stick together and distance themselves from it."[34] After a few days, five of the foreigners decided to leave and join another unit. "I'm more scared of them than the fucking Serbs," Hutt admitted.[35] As they were preparing to leave, the foreigners had a tense standoff with the HOS unit's criminal ringleaders. The latter made the foreigners leave their weapons behind. Walking away, Hutt and his friends were convinced that they would be shot in the back, but one of the Croats broke rank to escort them to safety.

The misconduct of locals also prompted Roland Bartetzko to leave the Bosnian-Croat HVO unit he served with in Mostar in April 1994. This happened after he witnessed "some very cruel scenes, which could be described as atrocities." Instead, he joined a platoon composed of foreigners, "which was much better in this regard, but lacked discipline."[36]

The foreign volunteers who have joined the jihadist group Al-Shabaab in Somalia since 2006—initially to fight against invading Ethiopian forces, and subsequently against African Union Mission in Somalia (AMISOM) peacekeeping troops from Uganda and Burundi—found themselves in a difficult predicament. For international audiences, the American volunteer Omar Shafik Hammami, also known as Abu Mansoor al-Amriki, became the face of Al-Shabaab. He was one of the first foreigners to join the organization and featured prominently in its propaganda. He posted clips on YouTube encouraging Muslims in the West to join the jihadi movement.[37] However, by 2013 an internal feud within Al-Shabaab began to claim the lives of prominent figures within the organization, including volunteers who had gone to Somalia from other countries. Hammami gave an interview in which he announced he was no longer part of Al-Shabaab. In April of that year he used Twitter to claim he had "just been shot in neck by shabab assassin," attaching a photograph showing a neck wound. Then, on 12 September 2013, he was killed. Even though he was wanted by the United States, his comrades believed the assassination was carried out by Al-Shabaab.[38]

In some cases, the repercussions of the choices that foreign volunteers made were only felt once they left the warzone. Born in Colorado Springs and raised in Arizona, Eric Harroun had served in the US Army and was discharged following a road accident–related head injury. Some ten years later, in early 2013, he traveled to Syria and joined the Free Syrian Army. Soon Harroun found himself involved in a combat operation in which a number of rebel groups tried, unsuccessfully, to storm a position held by troops loyal to the Assad regime near Idlib. Harroun hardly spoke any Arabic and found it difficult to follow what was going on. During the retreat, he jumped on the back of the wrong pickup truck. All he was able to gather was that his new companions were more religious than the Free Syrian Army soldiers he met previously and that the group he joined had a black flag. Harroun stayed with this group for a number of weeks, posting a few clips online showing the action he was witnessing. After about six weeks he left Syria and went to the American consulate in Istanbul. He was questioned, flown back to the United States, and eventually arrested. The authorities (and apparently Harroun himself) believed he had served with Al-Nusra Front, which by then was designated as a terrorist organization by the US government. The charges he initially faced included some punishable by death. Only after the videos he had participated in while in Syria were properly translated did it transpire that the organization he had joined was Al-Nasr, not Al-Nusra. His charges were thus reduced, and he was eventually released after striking a deal with the prosecution.[39]

As these examples illustrate, voluntary military service abroad often entails a great deal of uncertainty and has led to disappointment in a large number of cases. Let us now examine some of the longer-term effects that the volunteers' wartime experiences have had on their morale.

Disillusionment versus Radicalization

Foreign volunteers often arrive in their respective conflict zones with a set of preconceived ideas about why they are there, the people they have come to support, and the enemy they intend to fight against. Such notions seldom match the reality they encounter. As a result of this clash between expectations and the situation on the ground, many become disillusioned, a process to which either the misconduct of fellow volunteers or negative reception by their hosts (or both) often contributes.

At the same time, other volunteers are enthralled by the experience of foreign military service and consequently become more committed to the cause—or radicalized, in contemporary parlance—than they had been before they arrived. Radicalization and disillusionment are relative and, to a certain extent, subjective terms. To say that a person has been "radicalized" presupposes that she or he was not fully committed to the cause before plunging into action—in our case, voluntary military service abroad. Such a notion is, of course, very difficult to prove or disprove. Thus, for our discussion, we have selected examples where wartime service had a demonstrable impact on individuals who remained activists, in various capacities, for the same cause or a similar one after leaving the conflict zone.

Disillusioned and radicalized foreign volunteers often emerge from the same conflict. The ratio between these two groups cannot be determined with any degree of certainty because there is no objective way of counting "disillusioned" and "radicalized" veterans. Nonetheless, a few general observations are possible. In most historical cases the radicalized do not seem to have outnumbered those who became disillusioned, though the former have often been able to attract more attention due to their postwar activities. To redress the balance in exposure, let us begin with the examples of Greece in the 1820s and Serbia in 1876, which illustrate how transnational volunteering had led mainly to a loss of faith in the cause, before examining less clear-cut cases from later conflicts.

The disparity between expectations and reality was particularly stark for the philhellenes. We have already seen in Chapter 5 how the Greek Committee in Darmstadt began to organize the German Legion, the first contingent of which left Marseilles in November 1822. Its leader was a man called Kephalas, a Greek who had been living in western Europe and who managed to convince Mavrokordatos to support this endeavor. The brig *Scipio,* which ferried the volunteers, arrived in Hydra in December. Its reception by the locals was anything but cordial. The legionaries were not permitted to come ashore. One Swiss soldier who had become ill during the voyage died in the filthy and cramped conditions that prevailed on board the ship. Only then did the volunteers receive permission to disembark, in small parties, and on condition that they did not enter the town. Once ashore, they encountered two volunteers from Bremen who had arrived in Greece before them. The two lay in rags, covered in vermin and suffering from fever. One of them died within a few days. Back in the port, a mutiny broke out and Kephalas lost much of his authority. Eventually, Legion members agreed to move to Nauplia, where they hoped the Greek leadership would put them to good use. However, in Nauplia too there was nothing for them to do. Running out of money, they had to sell their possessions and their weapons. By the summer of 1823, the Legion had disintegrated. At least twenty-five died of disease, including Kephalas. Some of the volunteers tried to make their own way back home. Those who remained were said to be subsisting on tortoises.[40]

Meanwhile, the authorities in Marseilles—the departing point for most of the philhellenic volunteers—had been interviewing disgruntled veterans returning from the war in Greece since 1821. Deciding it was pointless to send young men to a fate of death or misery, they banned the dispatching of further contingents through the port in late 1822. The former volunteer and historian of the war in Greece Thomas Gordon later remarked, "The battle of Petta, and diseases incident to the climate, thinned their number, and many of the survivors left a country where their enthusiasm could not bear up against want and neglect."[41]

Disillusionment was also rife among Russian volunteers shortly after their arrival in Serbia in 1876. In early September the Russian consul in Belgrade, Andrei Nikolaevich Kartsov, wrote to his foreign minister, Nicholas Karlovich Giers, painting a gloomy picture of the Pan-Slavic endeavor.

> I am sorry for our youth which is involuntarily disillusioned here after two or three days in the Serbian army. Drawn by the desire to take part in a military affair, they see that they are not among orderly troops but in some caravan or group of boastful cowards; and those who come because of sympathy for the Slavonic brotherhood discover with bitter amazement to what extent his own skin is precious to his brother and how little he values the sacrifice and the very life of those of the same faith who arrive to spill blood for the Serbian cause.[42]

The leader of the Russian contingent, General Cherniaev, continuously bickered with the Serbian leadership over tactics and politics. When the Serb force at Đunis, which included the Russian contingent, was defeated by the Ottomans in late October, Cherniaev—according to a number of contemporary commentators—shared much of the blame. On the one hand, his staff was incompetent, and the general was unable to coordinate and discipline the large force at his disposal. On the other hand, the Serbs were displeased with Cherniaev's Russian-style frontal assaults. The Serbian war minister, Nikolić, argued that the general "does not spare men. It is possible to conduct war this way with Russian troops, whose losses can be replaced, but not with Serbian militia, since the entire country would soon go into mourning."[43] Nikolić's criticism was not restricted to Cherniaev. He pointed out that he had been receiving a stream of "complaints against the Russian volunteers who overindulge in spirits . . . and commit scandalous acts in hotels, cafes, and the streets."[44] Cherniaev left Serbia, discredited, in late November. Some of his men joined Serbian units, but most went back to Russia where they spoke disparagingly of their commander's collapse under the strain of battle.

While the philhellenic contingents of the early 1820s and the Pan-Slavic endeavor in Serbia in 1876 often resulted in the disillusionment of the volunteers, the service of foreigners in the Russian Civil War produced more complex results. The Bolshevik firebrand Leon Trotsky, who created the Red Army, relied heavily on non-Russian recruits during the early stages of the war, both because they had military experience and because they were perceived as loyal to the new regime. In this context, troops from Latvia, initially summoned to Petrograd to defend the nascent Bolshevik government in late 1917, soon acquired a reputation for being the regime's praetorian guard. Trotsky used two Latvian battalions to crush a Socialist Revolutionary Party (SR) insurrection against the Bolsheviks in Moscow in early July 1918. Latvian troops also fought against the Czechoslovak Legion, a force of former First World War prisoners of war that opposed the Bolsheviks. However, during the summer and autumn of 1918, the morale of the Latvians deteriorated as a result of being deployed to put down food riots and peasant disturbances. The regime's surveillance indicated that the riflemen's belief in the cause was starting to wane. Soon Latvian troops refused to fight. In September only 90 of the original 500 men of the Latvian 4th Regiment were still under arms. On their return from Russia, former riflemen did not support the short-lived Latvian Soviet Government, which existed between January and May 1919. According to historian Geoffrey Swain, the veterans' disillusionment did much to sap local support for the attempt to install a Soviet administration in Latvia.[45]

Béla Kun, a former Austro-Hungarian soldier who was captured by the Russians in 1916 and went on to join the Bolsheviks, was also in Moscow during the SR insurrection of July 1918. He led a group of about 150 Hungarian volunteers who, together with a Latvian company, stormed the central post office building that had been taken over by SR troops.[46] Kun returned to

Hungary in November 1918 and played a leading role in the attempt to establish the Hungarian Soviet Republic in 1919. After its collapse he moved to Moscow where he remained an active Communist until his arrest, trial, and execution in the late 1930s.[47] Historian Ivan Volgyes argues that Kun's zeal might not be representative of all the Hungarian former prisoners of war who fought alongside the Red Army in 1918. While some of them joined the Bolsheviks out of a spirit of revolution, others enlisted simply to improve their lot. It was the hostile and suspicious reception that awaited them upon their return to Hungary that pushed them to embrace Bolshevism as a solution for what they perceived as the ills of their home state.[48]

The story of Josip Broz shares a few similarities with Kun's. Born to a Croatian father and a Slovenian mother, he served as a noncommissioned officer in the Austro-Hungarian Army during the First World War. He was injured and taken prisoner by the Russians in 1915. After spending the next two years convalescing and, later, in a prisoner of war camp, he was able to break free following the revolution of February 1917, which toppled the tsar. His biographers differ on his level of ideological commitment at this stage. Some depict him as an already formed leftist revolutionary while others argue that his ideological zeal developed later.[49] In any case, it was at this point that Broz befriended a Bolshevik engineer and subsequently took part in Bolshevik demonstrations against the provisional government in the capital, Petrograd, before being arrested and dispatched to Siberia. He managed to escape from the train that ferried him eastward, making his own way to Omsk, where he learned that the Bolsheviks had seized power in what became known as the October Revolution. He then joined the hastily established Red Guard and spent much of the winter of 1917–1918 on sentry duty on the railway near Omsk. When the Whites captured the area, Broz went into hiding in a Kyrgyz village. He emerged in early 1920, after White forces had been driven out of Omsk, and married a young local Russian Bolshevik. Later that year he returned to his native Croatia, which had by then become part of the Kingdom of Serbs, Croats, and Slovenes. His military role in the Russian Civil War was negligible. However, the revolution and the war sharpened his ideological identity. Broz—better known by his revolutionary name, Tito—went on to play a key role in the Communist Party of Yugoslavia in the interwar period, as a partisan leader during the Second World War, and as Yugoslavia's dictator throughout much of the Cold War.

Like the civil war in Russia, the Spanish Civil War gave rise to cases of both radicalization and disillusionment. When the nurse Patience Darton went to Spain, she held left-leaning views but was not a member of the Communist Party. While in Spain, she fell in love with Robert Aaquist, a German Jew who arrived in Spain from Palestine where he had lived for a number of years. "The longer I am in a country at war the more militantly antifascist I become," Darton wrote to her lover in July 1938.[50] When Aaquist was killed soon afterward during the Battle of the Ebro, Darton made a symbolic gesture to prove her commitment to the cause by joining and working for the Communist Party. In the 1950s she moved to live in China, seeing a link between the Spanish and Chinese struggles of the 1930s.

Yugoslav volunteers in the International Brigades provide another example of how foreign wartime service could act as a catalyst for enhanced ideological dedication. The number of Yugoslavs who fought in Spain has been estimated at between 1,575 and 1,592.[51] Out of these, only about 560 were members of the Yugoslav Communist Party prior to their arrival in Spain. Between 35 and 50 percent of the Yugoslav contingent was killed during the conflict and a further 300 were wounded. In 1939, following the collapse of the Spanish Republic, a large number of former volunteers were interned in camps in southern France. About 250 to 300 veterans managed to return to Yugoslavia. They had to do so illegally because the government in Belgrade had stripped them of their Yugoslav citizenship. Even though most of the volunteers who went to Spain were not party members, the majority of those who returned identified or had come to identify with the Communist cause. Many of the returnees went on to play leading roles in Tito's partisan army during the Second World War. These individuals were valued not only for their military experience but also for their political loyalty. Veterans of Spain commanded the partisans' four field armies and, in all, 30 former International Brigaders attained the rank of general by 1945. After the Second World War several veterans held key governmental positions in the Socialist Federal Republic of Yugoslavia.[52]

Conversely, there were many cases of desertion by foreign volunteers during the Spanish Civil War, including from the ranks of the International Brigades. In October 1937, for instance, fourteen men of the 15th International Brigade were put on trial in Aragón for desertion. Two of the men arrested for desertion, both from the United States, had stolen an ambulance and planned to reach the French border. They were sentenced to death although it appears they escaped execution. The practice of executing—or at least decreeing the execution of—deserters, "wreckers," and "Trotskyites" in the International Brigades has spawned a historiographic debate about how Stalinist methods had been imported into Spain by Comintern officials and secret agents of the Soviet secret police, the NKVD.[53] There is some controversy around the actual number of International Brigaders who were executed under the orders of André Marty, the French Communist political commissar who ran the camp at Albacete.[54] There is little doubt, however, that some foreign volunteers became utterly disillusioned either with the realities of the battlefield, with the political feuds that the war created, or with both. The British and French consuls in Valencia and elsewhere in Republican Spain dealt with the repatriation of hundreds of deserters, especially after the bloody battles of 1937 and 1938.[55]

Those who deserted were not the only ones to feel disappointed to some extent with the war. Indeed, as George Orwell famously concluded, "I suppose there is no one who spent more than a few weeks in Spain without being in some degree disillusioned."[56] Jason Gurney, for instance, was disheartened by the passion of the communists in the International Brigades for "conspiratorial activity and its corollary of suspicion." He wrote of mysterious disappearances from the ranks and of secret trials for real or imagined offenses. He later lamented that "the nobility of the cause for which I had come to Spain was clearly a fiction."[57] Returning to Britain after sustaining a serious injury at the front, Gurney wrote a memoir that was critical of the way in which the International Brigades were "used to make a political point" rather than try and win the war.[58] Peter Kemp, who fought on the Nationalist side, realized in retrospect that "it is much easier to get into a war than out; also, more significantly, that in any war, what you think you're fighting for is not what emerges at the end, even from victory."[59]

Before assessing cases of radicalization among the foreigners who enlisted in the Waffen SS during the Second World War, it is useful to consider the related—though not identical—phenomenon of wartime service leading to the "brutalization" of soldiers. In his compelling analysis of the wartime conduct of Police Battalion 101 in Poland, historian Christopher Browning argues that a combination of factors turned the "ordinary men" of this German Order Police unit into killers. The polarizing environment created by the war in conjunction with insidious racist propaganda, indoctrination, and peer pressure to conform created the conditions whereby most of the middle-aged police reservists in the unit took an active part in the mass murder of Jewish civilians. Furthermore, Browning points out that the wholesale executions they committed were not the result of "battle-field frenzy" in which soldiers may become inured to violence and numbed to the taking of human lives.[60]

To what extent did foreign volunteers in the Waffen SS undergo a similar process of brutalization? Danish volunteers, for instance, were in general ideologically committed, perhaps more so than the reservists in Police Battalion 101. In the summer of 1941 and between October 1942 and June 1943, Waffen SS recruits from Denmark were present when war crimes against civilians were being committed on the eastern front, though it cannot be categorically proven that they took an active part in executions.[61] But while we cannot say with certainty that the conditions in which Danish volunteers served replicated those in the unit assessed by Browning, there is evidence to show that at least some of them continued the fight even when they left the eastern front. In the autumn of 1943, when several Danes from the Waffen SS were on leave in Copenhagen, they voluntarily took part in the failed attempt to round up Danish Jews for deportation (an attempt that was foiled by Danish resistance). One of the volunteers, Søren Kam, also kidnapped and murdered the Danish editor and suspected resistance member Carl Henrik Clemmensen. After being acquitted by an SS court in Berlin, he returned to active service on the eastern front.[62] Evidently, some of the foreigners in the Waffen SS internalized the Nazi ethos.

Yet the historiographic prominence of such cases of brutalization or radicalization must not deceive us into believing that this was the path taken by all the foreigners in the Waffen SS. As in previous historical examples, disillusionment was rife also among this cohort of volunteers. We have already seen how divisions that were raised by the Germans in the Balkans suffered from mutinies and desertion and were eventually dissolved. In some cases, volunteers from western Europe received harsh treatment from their German superiors and their morale deteriorated accordingly. By March 1942 Flemish volunteers complained that they had been beaten, threatened with pistols, and subjected to verbal abuse during their training period. The collaborationist Flemish leader Staf de Clercq wrote to Himmler saying that volunteers from among his countrymen were being called a "nation of idiots."[63] Moreover, he argued that many had been duped into military service. Of 875 men in the Flandern Legion, he said, 500 had been working in France and were induced by promises of higher pay to volunteer for labor in Poland. Only upon arrival did they discover that they had, in fact, joined the Waffen SS, much to their dissatisfaction. Even if we take de Clercq's statement that these men had been duped with a pinch of salt, it is clear that they were demoralized. Other volunteers, who enlisted for one year and served in the Wiking Division, were not demobilized on time. Some of these crossed the border into neutral Sweden while on leave. According to one report, by 1942 even supposedly ideologically devoted Norwegian Nazi volunteers were becoming disillusioned.[64]

There were also cases of disillusionment among the volunteers who went to Israel in 1948. In fact, some of the foreigners changed their minds about volunteering before even reaching Israel. For instance, 21-year-old Wallace Levey from London traveled to Paris in May. Having reached the French capital, the Zionist network ferried him and several other volunteers from their headquarters at the Avenue de la Grande Armée to the Marseilles area, where they were to wait until they could be transported to Israel. Levey received some basic training at a camp outside the village of Trets. His passport was taken away from him. Given another man's identity card, he was told to pretend to be a Polish displaced person. Later, when he was told to hand over his British money, he began to "feel that there was something 'fishy' about the whole business." After wounding his arm and going into Marseilles for medical treatment, he went to see the British vice consul, who arranged for him to travel to Paris and from there back to Britain. On his way he reportedly met six other British volunteers who decided to quit and head back home.[65] The living conditions in Grand Arenas, the central displaced persons' camp where most of the volunteers waited before traveling to Israel, were very difficult. Stanley Jackson, also from London, "got fed up with life in the camp" and, together with a friend, escaped from the camp by "scaling the barbed wire surrounding it." He too sought help from the British consulate in Marseilles, as his passport had been taken away from him.[66]

However, as with the previous conflicts, there were also instances of exemplary dedication by foreigners. In October 1948 the commander of the fledgling Israeli air force, Ahron Remez, was asked by the acting IDF chief of staff to send one of his bombers to carry out a high-risk, low-altitude raid on a seemingly impenetrable fortress in the Negev Desert that was held by the Egyptian army. Leonard Fitchett, a former Royal Canadian Air Force pilot from Victoria, British Columbia, forcefully volunteered for this mission, knowing full well how dangerous it would be. Remez later recalled asking him why he—a non-Jew—should take such incredible risks. Fitchett explained that while the local pilots had guts, they did not have his Second World War experience, and that a person without ideals for which he is prepared to die has nothing to live for. He managed to hit his target once, but his aircraft was shot down during a second attempt.[67]

Since the early 1990s the prospect of foreign volunteers becoming more radicalized during their wartime service abroad developed into a global concern. In this context, the advent of the "clash of civilizations" wave of fault-line conflicts brought with it an important novelty. The problem was no longer whether foreign volunteers who fought in one conflict would go on to take part in other wars; this had already been done by previous generations of transnational warriors. Instead, involvement in terrorism became the key concern. The first alarm bells were raised in the West when foreign veterans of the jihadist struggle in Afghanistan returned home to take part in attacks against the regimes in Egypt and Algeria in 1992–1995.[68] The blurring of the boundaries between foreign volunteers and terrorists was increased by Al-Qaeda. Its hardcore operatives, along with Osama bin Laden himself, were veterans of the war in Afghanistan. Bin Laden shifted the priority of the jihadi movement from the "near enemy" (pro-Western Arab regimes) to the "far enemy" (the United States). In February 1998 his organization issued a fatwa (ruling) that declared: "To kill Americans and their allies, both civil and military, is an individual duty of every Muslim."[69]

Considering the precedent set by those "Afghan Arabs" who became terrorists, Western leaders and security agencies have expressed concern about returning foreign fighters from the outset of the civil war in Syria. Former British prime minister Tony Blair, for instance, argued in June 2014 that "there is a real risk that Syria becomes a haven for terrorism worse than Afghanistan in the 1990s."[70] Analysts, meanwhile, have disagreed about the extent of the threat posed by foreign fighters returning to the West from that conflict. In an influential article published in February 2013 (before the Islamic State rose to prominence), the Norwegian expert on violent Islamism, Thomas

Hegghammer, distinguishes between home-grown terrorists and volunteers who travel abroad to fight for Islamist causes. He calculates that, between 1990 and 2010, only about one in nine foreign fighters became involved in terrorist attacks upon their return.[71] Writing a year later, Brookings analysts Daniel Byman and Jeremy Shapiro similarly caution against exaggerating the threat posed by foreign fighters returning from Syria and Iraq. Like Hegghammer, they recognize that attacks have taken place and are likely to reoccur. However, they argue that a number of mitigating factors reduce the threat: many of the foreign fighters die during the conflict; some go on to fight in other war zones, without returning home; and others become disillusioned with the cause. Another mitigating factor is the ability of intelligence services to thwart terrorist plots. While some attacks may succeed (from the terrorists' perspective), most of the foreign fighters are unlikely to engage in plots against the West.[72]

Terrorism expert Jytte Klausen sees the threat as much more serious. Working with a data set containing information on nearly 900 jihadists who were engaged in various conflicts abroad between 1993 and 2012 (excluding the Syrian Civil War), she calculates that about one in three of the returnees participated in terrorist plots once back in the West.[73] The difference between her findings and Hegghammer's can be partially explained by the fact that Klausen does not distinguish between foreign volunteers who fought overseas in jihadist insurgencies and foreign-trained terrorists who attempted to link up with Al-Qaeda-associated training camps abroad.

Terrorist attacks in Europe in 2014 and 2015 proved that former foreign fighters with military training can bring about deadlier results than amateur, home-grown terrorists. However, these tragic events do not support any clear-cut conclusions about the extent to which participation in the conflict in Syria radicalized the individuals responsible. At least six of the men who carried out the attacks in the Bataclan theatre and elsewhere in Paris on 13 November 2015 had been to Syria and joined the ranks of the Islamic State. However, it is difficult to say when precisely they became "radicalized." Belgian-born Abdelhamid Abaaoud, Brahim Abdeslam, and his brother, Salah, had all been involved in petty crime a few years earlier. Salah Abdeslam did not travel to Syria. According to friends and family members, he only stopped smoking and drinking and became more devout a couple of months before the attacks.[74]

The background of the assailants who carried out the attack on the headquarters of the magazine *Charlie Hebdo* in Paris in January 2015 further complicate any radicalization-centered analysis. None of them had served as foreign fighters abroad. One of them, French-born Chérif Kouachi, had been sentenced in 2008 for recruiting would-be volunteers to fight in Iraq and for intending to travel there himself. Meanwhile, Chérif's brother, Saïd, reportedly received military training from Al-Qaeda in Yemen.[75] There are many paths toward radicalization. Indeed, as a number of commentators have pointed out, the question of what came first, extremist individuals seeking an outlet in war and terrorism or the war in Syria that radicalized foreign volunteers from the West, is something of a chicken-and-egg debate.[76]

The Western-raised men and women that congregated in territories held by the Islamic State varied in their intentions and personalities. As far as intentions can be gauged from statements made to the media, they harbored different approaches toward their home states. Some of these foreign fighters, particularly those who went to Syria in the early stages of the conflict, did not embark on their journey because they hated the West. Rather, they went because "Muslims were being slaughtered" and they felt they "had to do something."[77] Others were more anti-Western but did not necessarily pose a direct threat. Speaking from Syria, Ifthekar Jaman, a 23-year-old volunteer from southern Britain, told the BBC in 2013 that "they [the British authorities] can rest assured I don't plan to go back—so it's not a problem for them."[78] And then there was Abdelhamid Abaaoud, believed to be the chief planner behind the Paris attacks of November 2015, who bragged in an interview with *Dabiq* about traveling to Europe "in order to terrorize the crusaders waging war against the Muslims."[79] Those foreign veterans of Syria who think like him are obviously substitute-conflict volunteers who ultimately seek to topple the existing order in their home state.

Islamic State foreign fighters also differed in their personalities. They included people who already had a violent history, like Mehdi Nemmouche, and others who may have been brutalized by the war. The future executioner Mohammed Emwazi, for instance, would have passed for an "ordinary man" during his studies at the University of Westminster in 2006–2009. It would be sheer speculation to say which, if any, dark thoughts lurked within his personality at that stage. On the other hand, some foreigners soon found that war was not to their liking. We have already seen how Ahmed Shurbaji decided to leave Syria after a relatively short stay and effectively turned himself in to the Israeli authorities, and how reports reached the West in December 2014 about the Islamic State executing foreigners who wanted to leave. Military service in the war in Syria often resulted not in radicalization but in disillusionment, although the "radicalized" tend to receive the lion's share of media attention. Length of stay in the war zone is a useful indicator of the commitment of individuals to the cause. The disillusioned are more likely to seek a way home quickly.

As the examples above from the nineteenth, twentieth, and twenty-first centuries illustrate, the dichotomy between disillusioned and, if not radicalized, then at least committed foreign volunteers can be discerned in practically every conflict. The trajectory that individuals have taken at the end of their military service abroad was determined by three factors: their personality and beliefs, the context they encountered in the war zone (in terms of their hosts and peers), and the wartime experiences they accumulated.

False Badge of Honor

Our discussion of the negative aspects of the foreign-volunteer phenomenon would not be complete without one more small, esoteric yet illuminating group. In January 1880 the notorious Australian bank robber and bushranger Andrew Scott, also known as "Captain Moonlite," stood trial in New South Wales. He was arrested after his gang held up a sheep station near Wagga Wagga where they had taken two children as hostages. During his trial Scott claimed that, in 1860, he had been one of the Red Shirt volunteers who fought with Garibaldi in Italy. Although he often sported a Red Shirt while breaking the law, historians doubt the credibility of this claim, arguing that Scott was probably seeking to bathe his activity in the reflected glory of anti-authoritarianism.[80]

In 1961 Israel Bar was a respected military analyst, contracted by the Israeli defense ministry to write a history of the war of 1948. Arriving in Palestine in 1938, he claimed to have been trained as an officer in his native Austria and to have obtained a PhD from the University of Vienna. Importantly for our purpose, he also claimed to have served as a volunteer in the International Brigades during the Spanish Civil War where, he said, he had been promoted to the rank of colonel. With such an impressive record, it was easy for him to join the pre-state Zionist paramilitary organization and, later, to serve as a lieutenant colonel in the IDF. Once his hopes for a promotion were dashed, he left the military in the early 1950s. However, he quickly established himself as a writer on military affairs and was appointed head of the department of military history at Tel Aviv University. He had close links with Ben Gurion and with the prime minister's protégé, Deputy Defense Minister Shimon Peres. Then, in March 1961, Bar was arrested and accused of being a Soviet spy.

During his interrogation he admitted to having invented several details in his biography. He had never been trained as an officer, nor received a PhD from the University of Vienna. He also never served in Spain, despite his phenomenal knowledge about the war, which included tactical details and the names and positions held by many actual volunteers. Moshe Dayan, who later served as Israel's defense minister in the late 1960s and early 1970s, recalled first meeting Bar when the latter came to visit the British-Jewish counterinsurgency unit in which Dayan served in the late 1930s. Bar "received a rifle," Dayan said, "and I noticed that he didn't even know how to hold a weapon. So how could he be a fighter and a colonel in Spain?"[81] Nonetheless, Bar was able to fool enough people for over twenty years to reach a position of considerable importance.

As the examples of Scott and Bar illustrate, being or appearing to be a foreign volunteer (even under false pretenses) could be beneficial in some circumstances. Although many actual volunteers were troublemakers, fell in with the wrong crowd, or became disillusioned, the aura that surrounds the phenomenon has endured. Indeed, some foreign volunteers like Byron, Garibaldi, Che Guevara, and Abdullah Azzam became the stuff of legends. Their memory, as the next chapter shows, lingered long after they themselves passed away.

* 8 *

Links in a Chain

Memory and Myth

Writing about the International Brigades of the Spanish Civil War, historian Michael W. Jackson describes the volunteers as "marginal men produced by economic and political upheavals of the time."[1] Does such a depiction, if we expand it to include women, adequately describe all the foreign volunteers examined in this book? Lord Byron was one of the most celebrated literary figures of his age. Mikhail Grigorevich Cherniaev was a well-known general in Russia before he embarked on his Balkan endeavor in 1876. Eoin O'Duffy was a prominent Irish politician when he led a legion of volunteers to Spain. By and large, however, examples of volunteers who left a leading position in their country's political, business, or cultural elite to fight for a foreign cause were few and far between. As we have seen, statistically, volunteers tended to be young men in their twenties and thirties. The overwhelming majority were, if not marginal, then certainly not well-known when they first volunteered for foreign military service. Yet, in quite a few cases, foreign volunteers went on to have very high-profile careers, sometimes directly aided by the exposure they had gained during their military service abroad. Giuseppe Garibaldi became Italy's towering national hero in the nineteenth century. Che Guevara rose to the level of an international revolutionary icon. And Osama bin Laden went on to become the best-known global terrorist of his age.

This chapter deals with how foreign volunteers are remembered. There have been a number of excellent studies about how the memory of national cohorts of volunteers have been shaped and reshaped through time in their home states. This literature has illustrated the ways in which popular perceptions of groups of former foreign volunteers have changed in accordance with broader cultural and political shifts that occurred in home-state societies. For instance, at the height of the Cold War, veterans of the International Brigades in the United States were suspect because of their links with communism. Meanwhile those who resided in East Germany were celebrated as members of a valiant ideological vanguard.[2] There have also been a number of recent attempts to assess how host societies and states remembered contingents of foreign volunteers who had fought on their behalf. These too have shown how the legacy foreign volunteers left behind them was recast according to changing historical contexts.[3] Our focus here is slightly different. We examine how foreign volunteers looked back and positioned themselves in relation to their predecessors—that is, individuals or groups of foreign volunteers from previous generations.

Many soldiers throughout history—both national and transnational—strove to commemorate and extol their wartime contribution. One way of doing so was by creating a link between present and historical causes. For instance, many twentieth-century soldiers presented themselves as fighting in a modern-day crusade. Dwight Eisenhower used the phrase "Crusade in Europe" to describe the Allied campaign against Nazi Germany in the Second World War. Conversely, General Franco's struggle against the Spanish Republic and Hitler's war against the Soviet Union were presented in terms of a crusade against Bolshevism. Foreign volunteers have also made use of the crusade analogy, even when they fought for very different, or even opposite, causes.[4] Such symbolic links with the past are a means of legitimizing and giving historical substance to present-day actions. They tell us much more about the self-perception of those that seek to establish the link than they do about the historical case that is being invoked.

There are a number of noteworthy patterns in the tendency of foreign volunteers to peg their cause to previous conflicts in which foreigners have fought. In some cases the link between one war and another was provided by a person or persons who participated in both. When Capt. Milton M. Rubenfeld, a former US Army Air Force pilot of Jewish descent, offered his services to the Jewish Agency in Palestine in December 1947, he explained: "In 1939 I enlisted in the Royal Air Force and fought for England because I felt I was helping the cause of the Jews. I desire to do the same thing now [for the Jews of Palestine]."[5] However, in many cases foreign volunteers and their supporters invoked a connection with the past without veterans providing the link or even giving their consent.

Historians like to speak of invented traditions, a selective process that strives to create a sense of continuity with a historically suitable past.[6] As the examples that follow illustrate, the appeal of some invented traditions of foreign volunteering are part and parcel of specific national movements. Such traditions could revolve around old debts of gratitude, which members of one nation theoretically owed those of another, or a sense that previous generations of foreign volunteers represented the positive aspects of their nation and therefore should be emulated. Other traditions of foreign volunteering resonate transnationally and are seized upon by volunteers coming from different countries. One thing that all the examples in this chapter share is the way in which the image associated with individuals or groups of foreign volunteers transcends the historic specificity of the conflicts they have fought in and contributes to the creation of enduring myths.

Historical Inspirations and Debts of Gratitude

The foreign heroes of the American Revolution have had a very long legacy. Several cities, towns, villages, and streets in the United States bear the name of the Frenchman Lafayette, for instance. Importantly for our purpose, the Marquis de Lafayette has served as a yardstick for a number of later foreign volunteers from across the world. Furthermore, his memory has instilled a sense of gratitude among some Americans who felt that their country had a moral debt toward France.

Comparisons between Lafayette and other foreign volunteers began during his lifetime. John Devereux was a Catholic Irishman who had been naturalized in the United States. In 1819 and early 1820 he was in Dublin to recruit an Irish Legion to fight alongside Bolívar in Spanish America. Although initially he had been an absentee general, dispatching his troops to Gran Colombia without joining them, he was received with great pomp and ceremony when he finally arrived there in 1821. There were some in the 1820s who even compared the recruitment attempts of Devereux to Lafayette's endeavors to bring French support to the War of Independence in North America. Meanwhile, Lafayette and Bolívar briefly corresponded with each other in 1825 and 1826, and the French hero sent the liberator of South America a number of gifts.[7] At the same time, Lafayette was passionate about the philhellenic cause. During his famous visit to the United States in 1824–1825, he campaigned for Greek independence. Lafayette was a source of inspiration for a number of Americans who traveled to Greece to fight against the Ottomans or to otherwise volunteer their services. The French hero even wrote a letter of introduction for one of these volunteers, Estwick Evans of New Hampshire.[8]

Lord Byron, the most famous foreign volunteer who went to Greece, was unfavorably compared to Lafayette by the American preacher, theologian, and literary critic Andrews Norton. Writing shortly after the poet's death, Norton was critical of Byron's morality and his extramarital affairs. He wrote disapprovingly of Byron's motivations for traveling to Greece: "weary of life, disgusted with his pursuits" but yet "desirous, as ever, of being distinguished by the admiration of the world." According to Norton, Byron was of little use to the Greeks militarily: "no one, we suppose, imagines that he rendered, or was capable of rendering, any important services to the cause of that country." But the difference between Lafayette and Byron was to be measured not only in terms of military contribution. Norton argued that distinguished foreign benefactors of a nation "have much influence upon the national character" of their hosts. He noted that "our own country [the United States] has, in that respect, been peculiarly fortunate." However, he believed that "it would have been unhappy for Greece, if Lord Byron had been her Lafayette."[9]

The memory of Lafayette could also be used to entice foreigners to volunteer. As we have seen, James W. Quiggle, an American diplomat in Antwerp, wrote to Giuseppe Garibaldi on 8 June 1861 upon hearing that the general was planning to "join the army of the North" in the American Civil War. Quiggle, who was very enthusiastic about enlisting the services of the Italian hero of two worlds, told Garibaldi that if he fought for the Union, "the name of La Fayette will not surpass yours."[10]

A new chapter in Lafayette's legacy was written during the First World War. In November 1914 Norman Prince of Massachusetts was studying to become a pilot. He discussed the idea of volunteering to fly for France with a fellow trainee, Frazier Curtis (the latter tried and failed to join the British Flying Corps). Traveling to France in January 1915, Price was able to enlist the support of several Americans residing in Paris. Robert Bliss, who worked at the American embassy in Paris, reached out on behalf of Prince to Jarousse de Sillac, an undersecretary at the French ministry of foreign affairs. Fortunately for Prince, the French official agreed to help. De Sillac wrote to the ministry of war, explaining the advantages for France in creating an American squadron:

> The United States would be proud of the fact that certain of her young men, acting as did Lafayette, have come to fight for France and civilization. The resulting sentiment of enthusiasm could have but one effect: to turn the Americans in the direction of the Allies. There is a precedent in the Legion of Garibaldi, which has had an undeniable good influence on Franco–Italian relations.[11]

The Lafayette Escadrille, N-124, and members of the subsequent Lafayette Flying Corps flew as part of the French war effort from spring 1916 to February 1918, when they were incorporated into the American Air Service.[12]

By 1940 the Lafayette Escadrille became a source of historical legitimacy in and of itself. During the early stages of the Second World War, when the RAF established the first of the Eagle Squadrons, it drew on the memory of the American airmen who flew for the Allies in the First World War. Group Captain Charles Sweeny, the Eagle Squadron's honorary commanding officer, offered what the British propagandists called a "link with the past." Sweeny had been one of the organizers of the group of volunteers from the United States who joined the French army in 1914, a group that eventually provided the core of the Lafayette Escadrille. His involvement in the establishment of the Eagle Squadron in autumn 1940 enabled *Flight* magazine to claim that "the new squadron is a successor to the famous Escadrille Lafayette of the [Great] war, and has a very strong link with that unit."[13] One may legitimately ask what a late eighteenth-century young French nobleman who went to North America to fight against the British had in common with the twentieth-century pilots from the United States who volunteered to defend Britain. The past, in this case, as in so many others, proved highly malleable.

Another foreign hero of the American War of Independence whose name continued to resonate until well after the First World War was Kosciuszko. The American pilots who fought for Poland during its war against the Soviet Union in 1919–1921 named their squadron after Kosciuszko. Captain Merian C. Cooper, who was the first to propose to the Poles the idea of setting up an American squadron, was partly motivated by memories that were passed down in his family. His great-great grandfather, Colonel John Cooper, served under General Casimir Pulaski, the Polish nobleman who fought for the independence of the United States and was one of the founders of the American cavalry in the 1770s. In 1919 Merian Cooper tried to convince the Polish leader Jozef Pilsudski that a squadron of experienced American airmen would not only bring military benefits but also had a sound moral justification. He told the marshal that he "wanted to act in Poland as Kosciuszko and Pulaski had acted in my own country."[14] This sentiment was shared by other members of the squadron. Summarizing its activities a few years later, the veteran Kenneth Malcolm Murray wrote:

> By the work of this little known flying legion, the pilots of the Kosciuszko Squadron thus repaid, in part, America's debt to Poland for the gallant service rendered, in 1776, by those great Polish patriots, Kazimir Pulaski and Tadeucz [sic] Kosciuszko, who gloriously bared their swords in the cause of our American independence.[15]

Much like the foreign heroes of the American Revolution, the figure of Lord Byron served as a model for future generations of foreign volunteers. Byron himself held his predecessor Kosciuszko in high esteem. His 1823 poem *The Age of Bronze* describes the name of the Polish hero as "that sound that crashes in the tyrant's ear."[16] Byron's death a year later inspired the young Boston physician Samuel Gridley Howe, fresh from Harvard's medical school, to travel to Missolonghi in January 1825. Howe went on to serve as both a military surgeon and a soldier. He completed his distinguished service as Surgeon-in-Chief of the Greek Fleet.[17]

Byron's legacy was especially potent in Britain. The poet was a source of inspiration for the British volunteers who joined the Philhellenic Legion, which tried to assist Greece in its war against the Ottomans in 1897.[18] Four decades later, during the Spanish Civil War, Byron was hailed as the starting point of a glorious tradition, which the men of the International Brigades continued and enlarged. Some of the British members of the International Brigades were dismissive of Byron's contribution to the Greek cause. But the link between Byron, the British (or "English") tradition, and the International Brigades proved stronger.[19] John Cornford, a Cambridge-born poet and communist who was killed in battle in December 1936, has been described by one historian as "the epitome of the heroic Byronic volunteer."[20] When the British Battalion's memorial souvenir was prepared after the war, its epigraph was a quote from Byron's *Childe Harold:*

> Yet, freedom yet, thy banner
> torn but flying
> Streams like a thundercloud
> against the wind[21]

The same words were used on a plaque commemorating members of the International Brigades from south Yorkshire, erected by the Sheffield City Council in 1986. If Byron was celebrated by small groups of veterans and their supporters in Britain, foreign volunteers from Ireland and Poland played a much more central role in their respective nationalist narratives.

Vanguard for the Nation

In terms of volunteering in foreign conflicts since the late eighteenth century, the Irish and the Poles shared a number of similarities. Neither Ireland nor Poland were independent until after the First World War. Both countries provided recruits in abundant numbers for the states that governed them. Irishmen filled the ranks of the British armed forces and served across the empire while Poles served in the armies of the partitioning powers, Russia, Prussia, and Austria. At the same time, both the Irish and the Polish nationalist mythologies attached great importance to detachments of exiles and émigrés who fought in foreign conflicts. From the Napoleonic Wars onward, Irish and Polish foreign volunteers were held up as romantic symbols of the national aspirations of their peoples. Even though the volunteers in question were motivated by diverse and complex reasons, their sacrifices were often interpreted along patriotic lines as an attempt to achieve recognition for their respective national causes. Finally, in both countries the practice of volunteering for foreign military service persisted until the Second World War.

The Polish legions that fought for the French between 1797 and 1801 sowed the seeds for an enduring myth and, as we have already seen, bequeathed to twentieth-century Poland its national anthem. The slogan "For our freedom and yours," a central tenet of Polish nationalism, encapsulates the Polish tradition of foreign volunteering throughout the nineteenth and the first half of the twentieth century. According to international relations theorist Karma Nabulsi, the motto originated from a proclamation Kosciuszko had made to the Russians: "Join your hearts with Poles, who fight for their freedom and yours."[22] During the Polish uprising of 1830–1831, the Polish historian and politician Joachim Lelewel turned the phrase into the motto "For our freedom and yours," which was painted on banners in Polish and in Russian.[23] After the uprising was crushed, scores of Polish nationalists and insurgents went into exile. Some of these went on to fight in and establish symbolic links between various European and extra-European conflicts.

Polish exiles participated in several insurrections across Europe in the revolutionary upheavals of 1848–1849. For instance, the poet Adam Mickiewicz, who had been visiting Rome when news of the revolutions began to spread, formed a legion composed of Polish volunteers so that it could assist in the liberation of Italy. He had long seen émigré Polish soldiers such as Dabrowski's legion, who fought for their freedom by fighting for the liberty of others, as epitomizing the very essence of Polishness. In the poet's eyes, they had "abandoned everything that was private" for the sake of "national tradition" and "sought their fatherland having no idea where to find it."[24] In May 1848 Mickiewicz submitted a proposal to the Provisional Government of Lombardy, outlining the Legion's mission as he saw it. His proposal included a clause about the Legion ceasing its service in the Italian army should it be summoned to serve in Poland. The legion raised by Mickiewicz fought against the Austrians in Lombardy in August 1848, against royalists in Genoa in April 1849, and elsewhere in Italy later that year.[25]

General Józef Bem, one of the heroes of the 1830–1831 revolt, was appointed by the Hungarian government as its military commander in late 1848. He subdued peasant uprisings in Transylvania, drove the Habsburg regulars out of the region, and even warded off a small Russian expeditionary force in early winter 1849. Bem's aide-de-camp in Hungary was the exiled Polish nobleman Antoni Aleksander Iliński, another veteran of the uprising of 1830–1831. Before serving in Hungary, Iliński fought in the Liberal Wars in Portugal (where Bem attempted to form a Polish Legion), in the First Carlist War in Spain, and for the French in Algeria. After Russian troops defeated the Hungarians in 1849, both Bem and Iliński fled to the Ottoman Empire. Iliński converted to Islam, joined the Turkish army, and became known as General Iskender Pasha. He fought against Russia in the Crimean War before dying in Istanbul in 1861. Polish exiles also fought for the Paris Commune in 1871. Finally, during the First World War, Poles fought for both the Allied and Central Powers.[26]

The ideal of Poles fighting for liberty in foreign conflicts endured, in a slightly different form, into the twentieth century, despite Poland regaining its independence at the end of the First World War. In the Spanish Civil War it was revived by communist volunteers who were hostile toward the authoritarian government in Warsaw. The names that were selected for units of the International Brigades where Poles were to serve illustrate how the Polish communists tried to appropriate traditional Polish values while also taking an openly hostile stance toward their government. The 13th International Brigade was named after Jaroslaw Dabrowski, a Polish officer who became one of the commanders of the Paris Commune. One of the battalions in this brigade was named after Mickiewicz. When a Polish-Jewish company was formed, it was named after Naftali Botwin, a Polish-Jewish communist who had been executed in Poland in 1925 for having assassinated a police informant. The Botwin company paraphrased the Poles' historical motto by using the words "For your freedom and ours" on their flag in Yiddish, Polish, and Spanish.[27]

In September 1939 Poland was once again divided, and its borders disappeared from the map of Europe. With German forces advancing from the west and Soviet troops advancing from the east, hundreds of thousands of Poles attempted to escape into exile. Among these exiles was a large portion of the personnel of Poland's air force. A Polish government-in-exile was formed in Paris in late September, and by June 1940 approximately 85,000 Poles fought to defend France when it too was invaded. After the fall of France, those Poles who were not killed in action or captured escaped to Britain. Once there, they were incorporated into the British war effort. By the end of the war some 17,000 Poles had served in the RAF, either as airmen or as ground crew. Among these were members of the Kosciuszko Squadron, which took part in the Battle of Britain in 1940. These Polish pilots flew in aircraft that bore the insignia designed by the American aviators who had established the squadron some twenty years earlier.[28]

In June 1946 members of the Kosciuszko Squadron were barred from participating in the victory parade in London because the British government did not want to antagonize Stalin. Once the Iron Curtain was lowered, Poland remained firmly within the Eastern Bloc. But the memory of Poland's foreign volunteers survived the Cold War. To mark Poland's accession to NATO, President Bill Clinton gave a speech in Warsaw on 9 June 1997 that tied together the historical debt of the United States to Poland and the motto that had symbolically linked generations of Polish foreign volunteers: "More than 200 years ago, you sent your sons to help to secure our future. America has never forgotten. Now together we will work to secure the future of an undivided Europe for your freedom and ours."[29]

* * *

The titles "Irish Brigade" and "Irish Legion" have been invoked in a number of conflicts and in different historical contexts. The roots of this tradition are to be found among the so-called Wild Geese who served in the armies of France, Spain, and Austria in the late seventeenth century and throughout the eighteenth century. In the mid-nineteenth century the nationalist Young Ireland movement began to establish a romantic myth around the Wild Geese. The mercenary elements surrounding their recruitment were discarded and their memory was molded into that of a heroic, nationalist vanguard. As an article in the Dublin weekly *The Nation* exclaimed in October 1843: "Our business is to re-enact at home the deeds that have made Ireland famous in every other clime."[30]

The long nineteenth century provided plenty of cases of Irishmen whose military service abroad could bolster the image of the Irish as courageous defenders of just causes. As we have seen in Chapter 1, the Irish regiments of the ancien régime in France were abolished in 1791. However, in 1803 the tradition was renewed when Napoleon created the Irish Legion. Some of the members of this legion were exiled veterans of the failed anti-British uprising of 1798. Their service in the Napoleonic Wars could be interpreted as an instance of substitute-conflict volunteering. Another Irish Legion, consisting of more than 1,700 members, traveled to Spanish America in 1819–1820 to fight for Bolívar. The Irish leader and champion of Catholic emancipation in Britain Daniel O'Connell gave his blessing to Devereux, the organizer and recruiter of the Legion. O'Connell hoped that the Legion would "give glory to Ireland." He also dispatched his son, Morgan, to join the Legion and sent a message to Bolívar in which he expressed the hope that "my son may be enabled to form one link in that kindly chain which, I hope, long binds in mutual affection the free people of Colombia and the gallant but unhappy natives of Ireland."[31]

Irishmen were prominent among the San Patricios or Saint Patrick's Battalion that was composed of soldiers of various nationalities who fought for Mexico during the US-Mexican War of 1846–1847. The nucleus of this battalion was formed by a band of Irish soldiers who defected from the ranks of the US Army and were led by the Galway-born artillery lieutenant John Riley. In 1860 approximately 1,800 Irish volunteers joined the Pontifical Zouaves to fight for Pope Pius IX. During the American Civil War there was an Irish Brigade in the Union army while some Irish immigrants also fought for the Confederates. In the Anglo-Boer War, a couple of small Irish Brigades were formed by Irishmen who resided in southern Africa. These legions, which also included men of various other European nationalities, fought against British forces. During the First World War Roger Casement, a British diplomat-turned-Irish nationalist, tried to recruit Irish prisoners of war that were captured by the Germans to join an Irish Brigade. Though it did not materialize, the goal of the brigade was to secure the freedom of Ireland through a German victory over Britain.[32]

As in Poland, foreign volunteering by Irish men (and women) continued even after independence. As we have seen, General O'Duffy established an Irish Legion that fought for Franco in the Spanish Civil War. Furthermore, tens of thousands of volunteers from the Irish Free State and its successor, Éire, served in the British armed forces during the interwar period and in the Second World War. The contribution made by Irish volunteers in Britain's time of need was such that certain members of the British elite felt compelled to offer them a gesture of appreciation. For instance, the First World War veteran Lieutenant General Hubert Gough wrote a letter to the *Times* calling for the formation of an Irish Brigade or an Irish Division. He mentioned the precedent set with the establishment of the American Eagle Squadrons in the RAF and spoke highly of the Irishmen's "excellent military qualities" and how "their long-cherished traditions of soldierly valour directly inspire them."[33]

However, others in Britain felt uncomfortable with the idea of forming an Irish Brigade precisely because of the history associated with this title. John Andrews, the prime minister of Northern Ireland, wrote to Churchill complaining that the Protestants of Ulster would object to joining an Irish Brigade whose name would be associated with previous Irish forces that had

> fought against England in the days of Marlborough, the Irish Brigade which fought against Britain in the Boer War, Sir Roger Casement's effort in the last war and finally with a body of "Blue Shirts" organized in Eire a few years ago to fight in the Spanish Civil War.[34]

Churchill was a descendant of the eighteenth-century Duke of Marlborough, John Churchill, who fought against Irish troops in the War of the Spanish Succession (1701–1714). However, in this case the British prime minister sided with the likes of Hubert Gough and supported the creation of the 38th (Irish) Brigade in early 1942. Far from shying away from using a historically contentious title, the British military leadership seems to have believed that the connection with previous legendary military formations would cultivate the esprit de corps of the Irish recruits.[35] Indeed, in both the Irish and the Polish cases, the mythology surrounding foreign volunteering served not only their respective national cause but also the military organizations that hosted them.

The Tradition of the Red Shirts

Giuseppe Garibaldi was not only a veteran of a number of foreign conflicts but also the father of a dynasty of foreign volunteers and the founder of a tradition that bore his name. This tradition was most potent in Italy, though its effects can also be traced transnationally. While Giuseppe was still alive, he was the one who decided which foreign causes his followers should endorse. For instance, in 1866 he gave his blessing to those Red Shirt volunteers who wanted to fight alongside the Christian Greek population in Crete who rebelled against the Ottoman Empire. Giuseppe had been a long-time supporter of the philhellenic cause, but the revolt in Crete caught him at a bad time. In the summer of 1866 Garibaldi's volunteers fought against Austrian troops in Tyrol as part of the Italian sideshow to the war between Austria and Prussia. They suffered heavy losses. Hence, only a few dozen Garibaldian volunteers (some of whom came from France rather than Italy) landed in Crete in November and December 1866. Giuseppe's son, Ricciotti, traveled to mainland Greece in March 1867 with a further group of Garibaldian volunteers. He planned to take part in the fighting but was soon convinced to return to Italy, where his father was planning to invade the Papal States and capture Rome by force.[36]

Both Menotti and Ricciotti Garibaldi joined their father, who led the Army of the Vosges in eastern France during the Franco-Prussian War of 1870–1871. Giuseppe Garibaldi also expressed his support for the Paris Commune in spring 1871.[37] Meanwhile, the Italian government sought to distance itself from the Garibaldis' actions. Diplomatic officials expressed concern over the republicanism of the Garibaldian volunteers, fearing that this force might be used against the Italian monarchy in the future.[38] Indeed, the Garibaldian ideal of volunteering to fight abroad for a just cause would remain the purview of socialists, anarchists, and republicans for many years.

In August 1875 a gathering of insurgents in Herzegovina sent an appeal to Giuseppe Garibaldi. The aging general, whose health was failing, issued two proclamations in October supporting the revolt against the Ottomans. He also gave his blessing to the international volunteers who went to fight in the conflict, most of whom were Italian and French.[39]

After Giuseppe's death, Ricciotti became the torch-bearer of the Garibaldian cause. He led a large Red Shirt contingent in the Greco-Turkish War of 1897. He and his followers depicted their military contribution to the conflict as part of a chain of mutual solidarity between Italian Red Shirts and the Greeks. This conflict also offered an early example of how the Garibaldian legacy could be appropriated by individuals with different agendas. A small contingent of foreign volunteers, who described themselves as socialists and were led by Enrico Bertet, wore red shirts and identified as *garibaldini* without seeking the family's consent. This greatly upset Ricciotti, who worked hard to hinder Bertet's efforts.[40] Although he was well into his sixties, Ricciotti also led an Italian volunteer legion that fought for the Greeks against the Turks in Epirus in 1912, during the First Balkan War.[41] By this stage, the foreign volunteering career of his eldest son, Giuseppe "Peppino" Garibaldi, was well on its way.

Peppino participated in wars on four continents. He was raised in and was well aware of the family tradition. "My father had always promised to take me along the next time he went to war," he wrote in his memoirs. As a teenager he was assured he "would be allowed to learn the art of war on battlefields instead of in barrack," adding that this too "was the Tradition."[42] He abandoned his engineering studies in Fermo in central Italy at the age of seventeen to join the international volunteers who had congregated in Athens to fight in the Greco-Turkish War of 1897. Peppino later recounted how his father was at first reluctant to accept him into his unit, though he eventually relented, making his son a corporal. Ricciotti explained that "for a Garibaldi there is no worthier post than either the supreme command or a simple corporalship."[43]

After the war in Greece, Peppino completed his engineering studies and traveled to South America. He planned to find a job there, but as he later remarked, "my real reason was to visit the battlefields of my grandfather where our Tradition was born."[44] While working in Uruguay he met an elderly man named Pedro who had fought alongside Giuseppe Garibaldi in the 1840s. Pedro gave Peppino all the reasons why his grandfather had selected red shirts for the uniform of his troops. The reason that seems to have resonated most with both men was "that a man in a red shirt can neither hide nor retreat."[45]

Peppino had heard of the exploits of Italian volunteers who were fighting alongside the Boers in their war against the British in southern Africa. He was particularly impressed by the Italian Corps, which was led by Captain Camillo Ricchiardi. The young Garibaldi wanted to join the Boer cause. Little did he know that his father would soon give him instructions to join British forces instead. When Ricciotti cabled his son concerning this conflict, Peppino was dismayed. Drawing on his family's tradition, he identified the republican, freedom-seeking Boers as the ones fighting for a just cause. Ricciotti saw things differently. He was critical of the abusive and racist manner in which the Boers treated the black population. He also reminded his son of the debt the Garibaldis, and Italy in general, owed to the British for their support during the wars of the Risorgimento. Peppino dutifully sailed to Cape Town to meet with the British commander, Lord Kitchener. With time, Peppino came around to his father's view. "Most of the Boers," he later wrote, "were religious bigots, who had robbed and enslaved the peaceful natives."[46] Assigned to an Australian cavalry unit—Peppino had been born in Australia— he took part in mopping up operations in the northern regions of the Orange Free State.[47]

After the Boer War, Peppino learned of a revolution that had broken out in Venezuela to depose the dictator Cipriano Castro.[48] Having convinced himself that Castro was a tyrant who oppressed liberty, had opponents tortured, and imposed heavy taxes, Garibaldi decided that "here was a fitting field of action for the serving of my tradition."[49] He volunteered his services and was put in command of the artillery of one of the rebel armies. Peppino fought in the Battle of Ciudad Bolivar in 1903 in which the rebels were defeated. He later returned to Europe and, by 1910, was in Mexico.

When the Mexican leader Francisco Madero launched his armed struggle against the federal president, Porfirio Díaz, that year, he believed that the revolution ought to be a national affair, free from foreign interference. He thus turned down a number of offers from foreigners who wanted to join rebel forces in northern Mexico. However, he was keen to accept the services of Peppino Garibaldi, who brought with him not only symbolic value, as a warrior for democracy and against oppression, but also vast military experience and technical know-how. Peppino explained to his host that "the Garibaldis regarded it as their duty to fight for the oppressed against tyranny," adding that he "wanted only a chance to do what my grandfather and father had done before me."[50] Peppino was given the rank of lieutenant colonel and, during the rebel campaign in northern Chihuahua in the early months of 1911, he led a force of Mexicans and approximately 100 men of diverse nationalities. He continued to act as Madero's military advisor throughout his tenure in Mexico. To commemorate his contribution to the revolution, Plaza Garibaldi in Mexico City was named after him.[51]

Peppino served alongside his father in the First Balkan War of 1912–1913. As we have seen, in the early months of the First World War he led the Garibaldian volunteers who joined the French Foreign Legion and fought against the Germans in Argonne, a battle where two members of the Garibaldi family died. After Italy joined the war, Peppino served in the Italian army, reaching the rank of general.

In the interwar years the Garibaldi legacy became contested, with both the Italian Fascists and the antifascist movement trying to appropriate its memory. One of Giuseppe Garibaldi's grandsons, Ezio, pledged his support to Mussolini, as did some of the veterans of the First World War Battle of Argonne. At the same time, many of the antifascist Italian volunteers who joined the International Brigades in the Spanish Civil War formed part of the Garibaldi Battalion. In March 1937 this battalion took part in the Battle of Guadalajara where Italian Fascists and antifascists fought against each other.[52] The battalion, together with other units, later formed part of the Garibaldi Brigade, which had its own newspaper, *Il Garibaldino*. Writing in July 1937, one contributor framed the Italian antifascist effort in Spain as part of a "glorious" patrimony of which "the proletariat must take care". "Today," he argued, "Garibaldi lives again" through the Brigade "which carries the name of the great hero" and seeks to emulate "the heroic deeds of our Risorgimento."[53]

The contested appropriation of Garibaldism did not end there. In the lead-up to and immediately after the outbreak of the Second World War in 1939, Sante Garibaldi, an antifascist exile and one of the grandchildren of Giuseppe Garibaldi, agreed in principle to lead an Italian Legion of volunteers who would fight for France. However, this proposal did not materialize because of the reluctance of the French government.[54] In the last two years of that war, communist partisan units in Italy named themselves after Garibaldi. More importantly for our purpose were the contemporaneous *garibaldini* across the Adriatic Sea. In autumn 1943, after the toppling of Mussolini's regime in Rome, approximately 16,000 members of two Italian divisions that were stationed in Montenegro formed the Italian Partisan Division "Garibaldi." The Yugoslav leader Tito incorporated this Garibaldi division into his partisan force.[55] Conversely, in his attempt to rehabilitate the Waffen SS after the Second World War, the former Nazi general Felix Steiner praised Giuseppe Garibaldi for his distinguished career and his followers for their unswerving commitment to their cause. He implied that the volunteers who joined the Waffen SS were in fact following the path laid down by Garibaldi and other nineteenth-century volunteer forces such as the philhellenes.[56]

For the present-day national association of Garibaldian veterans in Italy (Associazione Nazionale Veterani e Reduci Garibaldini), the Garibaldi division in Second World War Yugoslavia was the last accepted manifestation of the Red Shirt tradition.[57] However, in 1993, during the wars in former Yugoslavia, rumors circulated about a Garibaldi unit of Italian volunteers who allegedly fought alongside Croatian-Serb units.[58] That such reports are difficult to verify is a moot point. What is important is the enduring appeal of Garibaldi as a symbol for units of foreign volunteers, particularly of Italian origin. The use they made of Garibaldi's name could be read as an attempt to acquire legitimacy. Even though they fought in different contexts and for very different causes, Garibaldi units all sought to link themselves with the quintessential armed hero of Italian independence.

Transnational Appropriations of the International Brigades

The International Brigades of the Spanish Civil War were one, if not the, frame of reference in discussions surrounding many groups of foreign volunteers in subsequent conflicts. When the Winter War between the Soviet Union and Finland broke out in late 1939, the memory of the transnational volunteers who had fought in Spain was still very fresh. We have already seen how the British cabinet member Samuel Hoare was thinking in terms of the foreign involvement in the Spanish Civil War when he suggested to dispatch British volunteers to Finland. There were also direct links between the foreigners who fought for the Spanish Republic and those who enlisted to fight for the Finns. When the British trade unions leader Walter Citrine traveled to Finland during the war, to see firsthand the effects it had on Finnish society, he encountered a number of foreign volunteers from Scandinavia. One Swedish journalist told Citrine that there were a number of Swedish veterans of the International Brigades who went to Finland to fight "against Russian aggression." "They, at least, are consistent," Citrine replied. "They fought against aggression in Spain, and they are doing it here."[59]

A similar view was expressed by a small group of German and Austrian socialist veterans of the International Brigades who, after their retreat from Spain, were interned at the camp of Gurs, Ilôt B, in southwestern France. Following the Nazi-Soviet Pact of August 1939, which effectively divided Eastern Europe between Germany and the Soviet Union, Paul Eduard Koch, one of the internees, sought to disassociate his group from the communist veterans of the Spanish Civil War.[60] In January 1940 he wrote a letter to the *Picture Post,* a British magazine that had published an article titled "Should We Help Finland?" Koch explained that he and his comrades had gone to Spain "to defend for the Spanish people what we lost in our fatherland: the Democracy!" He went on to argue that "we have two mortal enemies: Hitler and Stalin." Thus, after the attack on Finland, "we stand without reservation on the side of the democratic nations."[61]

The precedent set by the thousands of foreign volunteers of the International Brigades was once again a subject of debate in 1948, during the first Arab-Israeli War, even though ideologically the two conflicts were quite different. Early that year Jules Cuburnek, a 30-year-old former captain in the US Army Air Force, was living in Chicago. A diaspora Jew whose Jewish identity was sharpened by his encounters with antisemitism during the Second World War, Cuburnek avidly followed reports about the events in Palestine. He later recalled having an argument with a fellow Second World War veteran about whether the Jews had kept "cushy" jobs during their military service, an accusation Cuburnek vehemently denied. He pointed out his own record, flying B-17 missions over Europe, and added that "there's no challenge in the world today. I wish there was something like the Spanish Civil War going on again where a guy could volunteer and do something for the world and change the world around."[62] Then one of the people present asked him, "Why don't you go to Palestine and fly for the Jews?" Cuburnek did not know that the Jews in Palestine had an air force, but two weeks later he volunteered. Soon he was part of a complex operation of smuggling aircraft from the United States through Panama, Dutch Guiana, Brazil, Dakar, Morocco, and Italy to the newly established state of Israel.[63]

The British volunteer pilot Gordon Levett believed that the conflicts in Spain and Palestine were somehow similar but felt that others did not recognize this similarity. In his memoirs, he lamented that, in 1948, "the Jewish cause did not inspire the support of the world's intellectuals as did, for example, the Spanish Civil War."[64] There were also direct links between the war in Spain and the Arab-Israeli War of 1948 in the form of volunteers who had served in both. One of them was Shmuel Segal. Born in Minsk, Segal grew up in Palestine where he joined the local Communist Youth Movement. In 1937 he traveled to France and volunteered to join the International Brigades in Spain. He fought in the Spanish Civil War through to the Battle of the Ebro. Once the International Brigades were dismantled and the foreign volunteers were withdrawn, Segal settled in Britain. He remained there until 1948, when he volunteered to fight for Israel and served as an artillery officer in the IDF.[65]

During the Cold War, the International Brigades were celebrated as a transnational manifestation of leftist solidarity in various contexts across the world.[66] In the mid-1980s Bill Bailey, a veteran of the Abraham Lincoln Brigade, praised the leftist volunteers from the United States who traveled to Nicaragua to support the Sandinista government. As a former self-appointed ambassador, he pointed out that in 1936 "the United States *did* turn its back on republican Spain by quickly enacting a nonintervention act, banning the sale of arms to either side." However, the foreign volunteers "were proving to the Spanish people that they were not alone in their struggle to safeguard their democracy" while also "telling the Spanish people by example that they did not agree with their government's position."[67] Bailey saw a clear parallel between the Spanish Civil War and the conflict in Nicaragua, and between the American volunteers who mobilized in both cases: "Like us, the men and women who volunteered to go to Spain and assist that democratic cause, they are eager to volunteer their services to the people of Nicaragua in whatever manner they know how."[68]

The International Brigades have also been appropriated in post–Cold War conflicts, even though the intellectual climate had changed markedly. Foreign volunteers such as Eduardo Rózsa-Flores and Gaston Besson, who came from leftist backgrounds and joined Croatian units during the Yugoslav Wars, spoke about the similarities between their contemporaries and the International Brigades of the 1930s.[69] However, many of the volunteers who fought alongside them, some of whom belonged to the far right, would have disagreed. Indeed, some saw themselves, instead, as continuing in the path of the Waffen SS.[70]

Finally, the memory of the International Brigades has also been invoked in the context of present-day Islamic foreign fighters. However, when such comparisons have been made, the aim was usually to highlight Western double standards. As Asad Khan, a trustee of the Convoy of Mercy charity organization, pointed out in 2001, "There are people going to all these places because they believe their fellow Muslims are suffering. The young men, who left Oxford and Cambridge in the 1930s to fight the Franco regime in Spain, were they terrorists?"[71]

Settling Scores

The enduring link between Russia and the Balkans, forged not only though state foreign policy but also by the military service of foreign volunteers, offers us a cautionary tale. We have already seen how, in 1876, General Cherniaev led thousands of Slavophile Russian volunteers who joined the Serbian army to fight against the Ottoman Empire. The sentiments expressed by some of the Russian volunteers who joined the armed forces of Republika Srpska in Bosnia in the 1990s were remarkably similar to those of their nineteenth-century predecessors. For instance, Igor Girkin, also known as Igor Ivanovich Strelkov, said he saw it as his duty to fight alongside his Orthodox co-religionists and against the Muslims, who had "dehumanized" the Serbs. According to a diary he published in 1999, Girkin served in Bosnia between the autumn of 1992 and the spring of 1993. Friends describe him as a person who longs for the return of the Russian monarchy. Having also served in the Russian military, Girkin emerged in 2014 as one of the leaders of the pro-Russian separatists in eastern Ukraine. In fact, he took on the title of minister of defense of the self-proclaimed Donetsk People's Republic.[72]

Reports about volunteers from Serbia who expressed a desire to fight alongside the Russians and against the Ukrainian government began to appear in the Balkan media almost as soon as the conflict in Crimea erupted in early 2014.[73] A handful of Serbian volunteers, led by Bratislav Živković, were filmed in uniform in Crimea in March of that year. They identified themselves as Chetniks, a name that harks back to the irregular bands (čete) that fought in the Balkan Wars of the 1910s, the monarchist and nationalist Serbian guerrilla movement of the Second World War, and nationalist paramilitary groups that took part in the Yugoslav Wars of the 1990s. Explaining his motivations, Živković spoke of his country's historical debt toward Russia: "Russians always came to volunteer in Serbian wars when we faced hardships and now we came here to help them."[74] When the conflict spread to eastern Ukraine, media reports began to emerge about other Serbian volunteers who fought on the pro-Russian separatist side. One of them was the 44-year-old Radomir Počuča, a former journalist, presenter on RTV Pink, and spokesman for the antiterrorism unit of the Serbian Interior Ministry. Počuča announced his presence in Ukraine through his Facebook profile. He said he had come to fight for the "holy Russian fatherland," adding that he felt he was among brothers.[75]

Here what seems like yet another example of volunteers feeling obliged to repay a perceived historical debt becomes more complicated. In early 2015 a group of about twenty volunteers from Croatia made their way to Ukraine to fight against the pro-Russian separatists. They pointed out that Ukraine had been one of the first countries to recognize Croatia's independence in December 1991.[76] In an interview given by two balaclava-clad volunteers, they explained that they were Croatian nationalists who support the maintenance of an independent Ukraine. They compared Ukraine's predicament vis-à-vis Russia with Serbia's aggression toward Croatia in 1991–1995. Well aware of the presence of some Chetniks fighting alongside the pro-Russian separatists, the Croatian volunteers sent the following sarcastic message:

> We know that some Serbian guys fight on the other side. [It] is true that "big Serbia" politics and "big Russia" politics, they are [the] same. . . . We all hope that we will meet somewhere our Serbian "friends" and want to say hello to them and hope to see you soon.[77]

In other words, Serbian and Croatian ultranationalists interpret their countries' history as well as contemporary international events in a way that allows them not only to repay perceived national debts of gratitude but also to confront supposed hereditary enemies on foreign battlefields.

＊　＊　＊

As the examples in this chapter illustrate, national and transnational mythologies about traditions of foreign volunteering or debts of gratitude, some of them dating back to the eighteenth or nineteenth century, still have contemporary resonance. A tradition of transnational volunteering is also invoked in the discourse that surrounds present-day Islamic foreign fighters. This tradition is fairly recent, dating back to the Afghan conflict against the Soviet Union in the 1980s. A few notable Afghan Arabs went on to create personal symbolic links between the war in Afghanistan and the conflicts that broke out in Bosnia, Tajikistan, and Chechnya in the 1990s. For instance, the Saudi-born veteran of Afghanistan Ibn Khattab fought in Tajikistan and later became a well-known commander in the First and Second Chechen Wars.[78]

Foreign volunteers have often been romanticized, especially in cases where the cause they fought for was widely perceived as being heroic or successful. For examples, foreigners who fought in Afghanistan are at the heart of the popular television miniseries *The Road to Kabul,* which was broadcast throughout the Arab world during Ramadan in 2004.[79] A romanticized image of past generations of foreign fighters could in theory feed into the motivations of prospective volunteers. However, contemporary foreign fighters—much like many of their non-Muslim predecessors—tend to exaggerate their links with previous generations of volunteers. They do so to increase their legitimacy and for propagandistic purposes. This is illustrated by the appropriation of the legacy of Abdullah Azzam, the architect of the recruitment of foreign volunteers for Afghanistan in the 1980s. A largely unknown organization calling itself the Abdullah Azzam Brigades first emerged in the early 2000s, when it claimed responsibility for attacks against tourists in Egypt. By the end of that decade the same name was being used by a more serious organization that had cells in a number of Middle Eastern countries. In November 2013 this eponymous group took responsibility for a suicide attack on the Iranian embassy in Beirut, Lebanon, which killed and injured scores of people. Had he been alive, would Abdullah Azzam have endorsed such attacks? The Algerian Abdullah Anas, who collaborated closely with Azzam in Pakistan in the 1980s, certainly did not think so. As he told *Time* in 2009:

> He [Azzam] called people to fight in Afghanistan because it was occupied by the Soviets. If he saw what happened in Iraq and what is happening in Palestine he would say the same thing. But what is going on in the name of jihad, killing civilians, kidnapping, hijacking airplanes, explosions in the public places—that is not what Abdullah Azzam called a jihad.[80]

Like Lafayette, Kosciuszko, Byron, Garibaldi, and other foreign volunteers before him, Azzam has a symbolic legacy associated with his name. Legacies such as these have a life of their own and are one of the reasons why foreign volunteers have a historical significance that goes beyond the actual military roles they played.

Epilogue

Foreign war volunteers are a modern phenomenon. Of course, the idea that people could travel far from their homes to join a war for the sake of something they believe in is an ancient one. What separates, say, the crusaders from post-eighteenth-century foreign volunteers is not so much the psychological mechanisms behind their decision making but the societal, cultural, and political contexts that surrounded them. It was only after states gained enough power to expect and demand their citizens to serve only in their "national" armed forces that the decision to volunteer for military service in a foreign conflict took on new meanings. Hence, modern-day foreign volunteering carries with it manifold political, diplomatic, and legal implications, even when individuals decide to enlist for highly personal reasons.

So what can we learn from almost two and a half centuries of foreign volunteering? Obviously, there are vast differences, for instance, in the varied causes foreigners fought for. Volunteers have been animated by different ideals and strove for different goals and outcomes in the conflicts they participated in. Some of them sought to help countries such as Gran Colombia (1816–1822), Greece (1821–1827), Israel (1947–1948), and Croatia (1991–1995) achieve independence. Others wanted to defend a sovereign country that had been invaded by a foreign power, as the examples of France (1870–1871, 1914),

Finland (1939–1940), and Afghanistan (1979–1989) illustrate. Others still wished to restore the status quo ante within a state, like the Spanish Carlists who fought to restore the Bourbons in southern Italy in the early 1860s or the International Brigades in the Spanish Civil War. And, finally, there were those who wanted to create new political entities as in the case of the foreigners who fought for the unification of Italy (1848–1849, 1860), Che Guevara in Bolivia (1967), or the contemporary foreign fighters who joined the Islamic State in Syria and Iraq.

Foreign volunteers also harbor different, and not always inimical, thoughts about their home state. Self-appointed ambassadors wish to see their home state pursue a different kind of foreign policy. Their volunteering may be an act of protest, but it is not necessarily an opening salvo for a revolution back home. When they decide to enlist, diaspora volunteers prioritize their historical homeland over their state of residence. At the end of their military engagement, volunteers from this category need to reconsider their priorities: some return to everything they have left behind in their state of residence while others seek to settle permanently in the historical homeland, especially in cases where the latter proved victorious. Cross-border volunteers want to enlarge their home state. They could prove troublesome for their government in those cases where they envisage internal political change to coincide with territorial expansion. However, governments should be most wary of substitute-conflict volunteers who invariably seek to overthrow the regimes in their home state.

The international landscape in which foreign volunteers have operated unavoidably changed over time. Not only has the intellectual climate shifted, bringing to the fore different ideologies, but also the borders on the world's map; the technology that volunteers used to communicate, travel, and fight; the legal framework with which they had to contend; and the surveillance to which they were subjected.

But there are also similarities between different generations of foreign volunteers. What generalizations can we make? To begin with, their gender breakdown is overwhelmingly masculine, with men making up between 90 and 100 percent of the foreign volunteers in each conflict. In terms of ages, there is a consistent pattern that runs from the Frenchmen and other Europeans who petitioned Benjamin Franklin for an opportunity to fight in the American Revolution in the 1770s to the Western volunteers who joined

Kurdish organizations in their fight against the Islamic State in the 2010s. The volunteers were people at different life stages, with some in their teens and others in their fifties or even sixties, yet the majority were men in their twenties and thirties. The median age varies from one conflict to another but is typically somewhere between 22 and 32.[1]

Whether they chose to fight for Israel or the Palestinian Arabs in 1948, the Afghani mujahideen in the 1980s or for other causes, those who volunteered their services emerged from a wider group of potential supporters. To use one contemporary example, the Facebook group "Lions of Rojava," which celebrates foreign volunteering for the Kurds in Syria, had 34,578 "likes" in late February 2017. Meanwhile, even the most generous estimates put the number of foreigners who have joined the YPG at several hundred.[2]

In terms of motivations, foreign volunteers respond to different push and pull factors that have changed over time. However, a search for meaning invariably underlined their actions. They could be wealthy and pay their own way or poor and depend on their recruiters for transportation and subsistence. They could be highly intelligent and proficient or on the opposite end of the spectrum. But, with a clear tendency toward restlessness, volunteers have consistently been proponents of a *vita activa*.[3]

Foreign volunteers also tend to have a blinkered view of, and superimpose their beliefs and perceptions on, the conflicts they join. Such perceptions are often at odds with the reality they encounter when they first arrive. On the battlefield, the performance of foreign volunteers has been mixed. Some of them exhibited extraordinary commitment and / or helped to professionalize the forces they joined, while others proved to be a burden on their hosts. Take, for example, Michael Enright, who was mentioned in the Introduction. The Hollywood actor, who joined the Kurdish YPG in Syria, was described by fellow Western foreign volunteers who served alongside him as "a liability" and "not just a danger to himself, he is a danger to everybody out here."[4] Even in cases where foreign volunteers were highly competent militarily, they were often hindered by language difficulties and other problems of cultural adaptation. Military effectiveness aside, foreign volunteers have been much more consistent in terms of their propagandistic value. The participation of foreigners in armed struggles abroad has been instrumental in raising international interest in such conflicts on numerous occasions.

One of the methods of getting global attention and, at the same time, increasing the appeal and legitimacy of the cause in question is by invoking links with the past. As we have seen, many foreign volunteers like to be associated with prominent predecessors, a practice that is still pertinent today. For instance, the 20-year-old British volunteer Ryan Lock, who joined the YPG and was killed in battle in December 2016, was given the Kurdish nom de guerre Berxwedan Givara, meaning "resistance Guevara," after the Argentinian-born revolutionary.[5]

The shared characteristics and commonalities across the generations discussed here clearly illustrate that foreign war volunteering constitutes a distinct diachronic phenomenon. Armed foreign volunteers share some motivations with international aid workers, on the one hand, and with mercenaries or military contractors, on the other. There have also been individuals who have crossed over between these categories. But, at the same time, the phenomenon has unique and persistent features that make it worthy of further study.

Contemporary foreign fighters are often depicted by world leaders and the media as a problem. The participation of a small number of former volunteers in terrorist attacks after returning from the Middle East contributes to their negative image. In May 2017, US Secretary of Defense James Mattis went as far as to argue that, in the campaign in Syria and Iraq against the Islamic State, "Our intention is that the foreign fighters do not survive the fight to return home to North Africa, to Europe, to America, to Asia, to Africa."[6] This heightened sense of threat was certainly not associated with all of the foreign volunteers examined in this book. The historical examples illustrate that public perception is cause dependent. Volunteers who fought for "good" or "just" causes tended to receive favorable coverage from like-minded media outlets and to enjoy a degree of popularity, whereas those who joined "bad" or "unjust" causes were perceived negatively.

A survey that was conducted by my students and me in 2016 confirms this tendency. In this study, 120 people aged 18 to 70 from across Britain were asked to describe their responses to four newspaper headlines reporting the death overseas of a member of the armed forces, a businessman (control question), a pro-Kurdish foreign volunteer, and an Islamic State foreign volunteer. The headlines specified that all the persons who died were from the United Kingdom. The British armed serviceman elicited the greatest amount of sympathy, with 83 percent of the respondents describing themselves as either sad or very sad. Nearly as many, 75 percent, were saddened or very saddened by both the death of the British businessman *and* the British volunteer who had been fighting alongside Kurdish militia. This result illustrates that while national legislation and international law may frown upon the phenomenon of foreign volunteers, the population at large does not necessarily agree, so long as the cause in question is perceived positively. The responses to the death of a British Islamic State volunteer were, perhaps unsurprisingly, markedly different: 55 percent of the respondents stated that they were indifferent to the headline while nearly 18 percent said that it made them either happy or very happy. Foreign volunteers, it seems, have the ability to touch a raw nerve.

Herein lies the allure of the phenomenon. Foreign volunteers make an exceptional decision that few others have. They also violate what has become an international norm: the expectation that if a person enlists for military service, he or she should do so within the armed forces of their own state. By traveling to take part in a conflict abroad, they create links between Poland and the United States, between the mining regions of Wales and the Second Spanish Republic, and between twenty-first century west London and al-Raqqah in Syria. They internationalize conflicts that occur in one region of the world and, at the same time, make these foreign conflicts matter back at home. In some cases they inspire romantic myths that have a very long afterlife, even though in reality foreign volunteers very often become disillusioned. When looking back at their historical predecessors, foreign volunteers like to see themselves as part of a chain of committed, brave, and selfless individuals. This book has tried to present them in all their complexity.

Notes

Introduction

1. Martin Robinson, Peter Allen, Louise Eccles, and David Williams, "'I Know ISIS Butcher Known as John the Jailer': French Former Hostage Says He Has a 'Rough' Idea Who Masked British Jihadi Is," *Daily Mail Online,* 21 August 2016, http://www.dailymail.co.uk/news/article-2730336/Find-British-butcher-mask -Jihadi-filmed-beheading-American-journalist-identified-John-Londoner-gang -UK-extremists-known-The-Beatles.html.

2. "Islamic State Militant 'Jihadi John' Named," *BBC,* 26 February 2015, http://www .bbc.co.uk/news/uk-31636419.

3. Robert Verkaik, "Mohammed Emwazi: 'Jihadi John' Warned Younger Brother Not to Follow Him to Syria and Isis," *Independent,* 24 January 2016, http://www .independent.co.uk/news/uk/home-news/jihadi-john-mohammed-emwazi -isis-syria-warned-brother-a6831666.html.

4. "Among the Believers There Are Men: Abu Muharib Al-Muhajir," *Dabiq* 13 (January 2016): 23.

5. An excerpt from the interview Enright gave to Al-Aan TV is featured in "From *Pirates of the Caribbean* to Fighting Daesh in Syria," *NRG,* 3 June 2015, http://www .nrg.co.il/online/47/ART2/698/499.html?hp=47&cat=308&loc=7.

6. Tony Blair, "Iraq, Syria and the Middle East," Office of Tony Blair, 14 June 2014, http://www.tonyblairoffice.org/news/entry/iraq-syria-and-the-middle-east-an -essay-by-tony-blair/.

7. Vladimir Putin, "A Plea for Caution from Russia," *New York Times,* 11 September 2013, http://www.nytimes.com/2013/09/12/opinion/putin-plea-for-caution-from-russia.html?_r=0.

8. Shaun Walker, "Russian Media Turn Attention to Syria as Ukraine Conflict Dies Down," *Guardian,* 1 October 2015, https://www.theguardian.com/world/2015/oct/01/russian-media-syria-ukraine?CMP=Share_iOSApp_Other.

9. Heather Saul, "Chinese Uighurs 'Join Isis Overseas and Return to Take Part in Terror Plots,' Officials Claim," *Independent,* 11 March 2015, http://www.independent.co.uk/news/world/asia/chinese-uighurs-join-isis-overseas-and-return-to-take-part-in-terror-plots-officials-claim-10099674.html.

10. "Security Council Unanimously Adopts Resolution Condemning Violent Extremism, Underscoring Need to Prevent Travel, Support for Foreign Terrorist Fighters," United Nations Meetings Coverage and Press Releases, 24 September 2015, http://www.un.org/press/en/2014/sc11580.doc.htm.

11. "Security Council Unanimously Adopts Resolution."

12. My approach is influenced by Stathis Kalyvas, *The Logic of Violence in Civil War* (New York, 2006), 10.

13. For alternative definitions, see Gilles Pécout, "The International Armed Volunteers: Pilgrims of a Transnational Risorgimento," *Journal of Modern Italian Studies* 14, no. 4 (2009): 414; and Andrea de Guttry, Francesca Capone, and Christophe Paulussen, eds., *Foreign Fighters under International Law and Beyond* (The Hague, 2016), 2.

14. The international law expert Ian Brownlie used the term "pseudo-volunteers" to describe such groups. Ian Brownlie, "Volunteers and the Law of War and Neutrality," *International and Comparative Law Quarterly* 5, no. 4 (1956): 570.

15. John F. Coverdale, *Italian Intervention in the Spanish Civil War* (Princeton, NJ, 1975), esp. 165–171, 393–396.

16. Xosé-Manoel Núñez Seixas, "An Approach to the Social Profile and the Ideological Motivations of the Spanish Volunteers of the 'Blue Division,' 1941–44," in *War Volunteering in Modern Times: From the French Revolution to the Second World War,* ed. Christine G. Krüger and Sonja Levsen (Basingstoke, 2011), 248–274; and Xavier Moreno Juliá, *The Blue Division: Spanish Blood in Russia, 1941–1945* (Brighton, 2015), esp. 293–312.

17. Robert Dallek, *Franklin D. Roosevelt and American Foreign Policy, 1932–1945* (New York, 1995), 273; and Jack Carpenter, "Flying Tigers," *New York Times,* 28 December 2003, 8.

18. John Quincy Adams, *Lives of Celebrated Statesmen* (New York, 1846), 45–53.

19. Thomas Gordon, *History of the Greek Revolution,* vol. 1 (Edinburgh, 1832), 295.

20. Maffeo Charles Poinsot, *Les volontaires étrangers de 1914* (Paris, 1915).

21. Felix Steiner, *Die Freiwilligen der Waffen-SS: Idee und Opfergang* (Oldendurf, 1973 [1958]).

22. David Malet, *Foreign Fighters: Transnational Identity in Civil Conflicts* (Oxford, 2013), 215–216.

23. Flora Sandes, *The Autobiography of a Woman Soldier: A Brief Record of Adventure with the Serbian Army, 1916–1919* (New York, 1927), 12–13.

24. Peter L. Bergen, *The Osama bin Laden I Know: An Oral History of al-Qaeda's Leader* (New York, 2006), 24–25, 41; and Darryl Li, "'Afghan Arabs,' Real and Imagined," *Middle East Report* 260 (2011), http://www.merip.org/mer/mer260/afghan-arabs -real-imagined.

25. *Heart of the PKK*, documentary, Journeyman Pictures, 2007, https://www .journeyman.tv/film/3686.

26. United Nations General Assembly, A / RES / 44 / 33, "International Convention against the Recruitment, Use, Financing and Training of Mercenaries," 4 December 1989, http://www.un.org/documents/ga/res/44/a44r034.htm.

27. See, for instance, Christopher Kinsey, "International Law and the Control of Mercenaries and Private Military Companies," *Cultures & Conflicts,* 26 June 2008, https://conflits.revues.org/11502.

28. Matthew Brown, *Adventuring through Spanish Colonies: Simón Bolívar, Foreign Mercenaries and the Birth of New Nations* (Liverpool, 2006); and Moises Enrique Rodríguez, *Freedom's Mercenaries: British Volunteers in the Wars of Independence of Latin America* (Lanham, MD, 2006).

29. Mike Hoare, *Congo Mercenary* (London, 1967).

30. Christian Koller, "Recruitment Policies and Recruitment Experiences in the French Foreign Legion," in *Transnational Soldiers: Foreign Military Enlistment in the Modern Era,* ed. Nir Arielli and Bruce Collins (Basingstoke, 2012), 87–104.

31. Gino Coletti, *Peppino Garibaldi e la legione garibaldina* (Bologna, 1915), 19; emphasis in the original.

32. "Mr. John Oliver Scarrott" (obituary), *Isle of Wight County Press,* 8 December 2015, http://www.iwcp.co.uk/news/obituaries/mr-john-oliver-scarrott-93370.aspx; and the author's correspondence with Antaine Mac Coscair (who fought alongside Scarrott in Croatia), January 2016.

33. Rob Krott, *Save the Last Bullet for Yourself: A Soldier of Fortune in the Balkans and Somalia* (Drexel Hill, PA, 2008).

34. Kevin Grant, Philippa Levine, and Frank Trentmann, *Beyond Empire: Britain, Empire and Transnationalism, c. 1880–1950* (Basingstoke, 2007), 2.

35. Max Weber, "Politics as a Vocation," in *Essay in Sociology* (Abingdon, 2005), 78.

36. Instances of foreign volunteering in East and Southeast Asia have received less attention. For more on transnational military service in these regions see, for

instance, Donald G. Gillin and Charles Etter, "Staying On: Japanese Soldiers and Civilians in China, 1945–1949," *Journal of Asian Studies*, 42, no. 3 (1983): 497–518; and Nicholas Farrelly, "Transnational Flows of Military Talent: The Contrasting Experiences of Burma and Thailand since the 1940s," in Arielli and Collins, *Transnational Soldiers*, 145–159.

1. Only a Nation in Arms?

1. Roger Williams, *The Actions of the Low Countries,* ed. D. W. Davies (New York, 1964 [1618]), 112–113.
2. D. W. Davies, Introduction to Williams, *The Actions of the Low Countries,* ix–xi.
3. Sir Roger Williams, *A Brief Discourse of Warre* (London, 1590), 3, 30.
4. Davies, Introduction, xxxix. Williams was nonetheless able to become a close advisor for the Dutch leader, William of Orange. According to one version, Williams was the one to capture the assassin of the Prince of Orange in July 1584. See D. J. B. Trim, "Williams, Sir Roger," in *Oxford Dictionary of National Biography* (Oxford, 2009), http://www.oxforddnb.com/index/101029543/Roger-Williams.
5. The notion of "feudal society" has come under challenge in recent decades, with historians conceding more and more that classic "feudalism" was only one part of political relations in the so-called feudal period. For a succinct discussion of the historiographic debate, see Richard Abels, "The Historiography of a Construct: 'Feudalism' and the Medieval Historian," *History Compass* 7, no. 2 (2009): 1008–1031.
6. Bernard S. Bachrach, "Medieval Military Historiography," in *Companion to Historiography,* ed. Michael Bentley (London, 1997), 204; Michael Howard, *War in European History* (Oxford, 2001), 8; and John Childs, *Armies and Warfare in Europe 1648–1789* (Manchester, 1982), 5–6.
7. Howard, *War in European History*, 25. See also Maria Nadia Covini, "Political and Military Bonds in the Italian State System," in *War and Competition between States,* ed. Philippe Contamine (Oxford, 2000), 19–24.
8. Niccolò Machiavelli, *History of Florence and of the Affairs of Italy, Book V* (London, 1847 [1521–1525]), 204.
9. Jan Lindegren, "Men, Money, and Means," in Contamine, *War and Competition between States,* 129–162, esp. 130; John Gooch, *Armies in Europe* (London, 1980), 7–9; Howard, *War in European History,* 22; and Hew Strachan, *European Armies and the Conduct of War* (London, 1992), 4.
10. Cited in Geoffrey Parker, *The Military Revolution* (Cambridge, 1996), 47.
11. Cited in Parker, *The Military Revolution,* 48. See also Thomas A. Brady, *German Histories in the Age of Reformations, 1400–1650* (Cambridge, 2009), 384.

12. Lothar Höbelt, "Surrender in the Thirty Year War," in *How Fighting Ends: A History of Surrender,* ed. Holger Afflerbach and Hew Strachan (Oxford, 2012), 141–152.

13. See, for instance, Ilya Berkovich, *Motivation in War: The Experience of Common Soldiers in Old-Regime Europe* (Cambridge, 2017), esp. 229–231.

14. Childs, *Armies and Warfare,* 8.

15. Howard, *War in European History,* 24. The sociologist Charles Tilly expressed a similar view in *Coercion, Capital, and European States, AD 990–1990* (Cambridge, MA, 1990), 81.

16. Parker, *The Military Revolution,* 60.

17. Parker, *The Military Revolution,* 62; Childs, *Armies and Warfare,* 42. For a more recent interpretation of the reasons behind the growth of European armies in this period, see Azar Gat, "What Constituted the Military Revolution of the Early Modern Period?," in *War in the Age of Revolution, 1775–1815,* ed. Roger Chickering and Stig Förster (Cambridge, 2010), 21–48.

18. Berkovich, *Motivation in War,* 4.

19. Cited in Eliot A. Cohen, *Citizens and Soldiers* (Ithaca, NY, 1985), 45–46. See also A. M. Nikolaieff, "Universal Military Service in Russia and Western Europe," *Russian Review* 8, no. 2 (1949): 117–118; Childs, *Armies and Warfare,* 45; and Peter Way, "The Scum of Every County, the Refuse of Mankind: Recruiting the British Army in the Eighteenth Century," in *Fighting for a Living: A Comparative Study of Military Labour,* ed. Erik-Jan Zürcher (Amsterdam, 2013), 291–330.

20. Strachan, *European Armies,* 9; Gooch, *Armies in Europe,* 11.

21. Cited in Strachan, *European Armies,* 9. See also Theodore Ropp, *War in the Modern World* (Baltimore, 2000), 53; Childs, *Armies and Warfare,* 47, 98–99; Nicholas Charles Pappas, *Greeks in Russian Military Service in the Eighteenth and Early Nineteenth Centuries* (Thessaloniki, 1991), 63; Janice E. Thomson, *Mercenaries, Pirates and Sovereigns: State-Building and Extraterritorial Violence in Early Modern Europe* (Princeton, NJ, 1994), 29; and Christopher J. Tozzi, *Nationalizing France's Army: Foreign, Black and Jewish Troops in the French Military, 1715–1831* (Charlottesville, VA, 2016), 19, 35.

22. Voltaire, *Candide, ou l'optimisme* (London, 1759), 8–10.

23. Cited in Thomas Hippler, "Heroism and the Nation during the French Revolutionary and Napoleonic Wars and the Age of Military Reform in Europe," in *Heroism & the Changing Character of War,* ed. Sibylle Scheipers (Basingstoke, 2014), 29. See also Strachan, *European Armies,* 25.

24. Daniel Krebs, "Desperate for Soldiers: The Recruitment of German Prisoners of War during the American War of Independence, 1776–1783," in *Transnational Soldiers: Foreign Military Enlistment in the Modern Era,* ed. Nir Arielli and Bruce

Collins (Basingstoke, 2012), 15–16; and "The Declaration of Independence: A Transcript," National Archives and Records Administration, http://www .archives.gov/exhibits/charters/declaration_transcript.html.

25. Tozzi, *Nationalizing France's Army*, 3, 6.

26. Alan Forrest, *Conscripts and Deserters: The Army and French Society during the Revolution and Empire* (New York, 1989), 16–26; John A. Lynn, "Toward an Army of Honor: The Moral Evolution of the French Army, 1789–1815," *French Historical Studies* 16, no. 1 (1989): 152–173; Sam Scott, "The French Revolution and the Irish Regiments in France," in *Ireland and the French Revolution,* ed. Hugh Gough and David Dickson (Dublin, 1990), 14–27; and Michael Rapport, *Nationality and Citizenship in Revolutionary France* (Oxford, 2000), 93.

27. Tozzi, *Nationalizing France's Army*, 7.

28. For more on the early history of the *Marseillaise,* see Julien Tiersot and O. T. Kindler, "Historic and National Songs of France," *Musical Quarterly* 6, no. 4 (1920): 599–632.

29. Rapport, *Nationality and Citizenship*, 154–155.

30. Cited in J. M. Thompson, *The French Revolution* (Oxford, 1966), 424.

31. Carl von Clausewitz, *On War,* ed. Michael Howard and Peter Paret (Princeton, NJ, 1989), 591.

32. Clausewitz, *On War*, 589, 591–92.

33. Clausewitz, *On War*, 220.

34. See, for instance, Howard, *War in European History*, 80, 96, 110; Geoffrey Best, *War and Society in Revolutionary Europe 1770–1870* (London, 1982), 63; Strachan, *European Armies,* 4, 40; George Mosse, *Fallen Soldiers: Reshaping the Memories of the World Wars* (New York, 1990), 15–19; Tilly, *Coercion,* 82–83; Russell F. Weigley, *The Age of Battles: The Quest for Decisive Warfare from Breitenfeld to Waterloo* (Bloomington, IN, 1991), 290; and John A. Lynn, "The Evolution of Army Style in the Modern West, 800–2000," *International History Review* 18, no. 3 (1996): 514.

35. Mosse, *Fallen Soldiers,* 18–19, 44, 50, 80, 92–93. For more on the symbolic importance of soldiers after the revolution, see Norman Hampson, "The French Revolution and the Nationalization of Honour," in *War and Society: Historical Essays in Honour and Memory of J. R. Western, 1928–1971,* ed. M. R. D. Foot (London, 1973), 199–212; Marie-Cecile Thoral, *From Valmy to Waterloo* (Basingstoke, 2011), esp. 20–31 and 72–80; Joseph Clarke, "'Valour Knows Neither Age Nor Sex': The *Recueil des Actions Héroïques* and the Representation of Courage in Revolutionary France," *War in History* 20, no. 1 (2013): 50–75; and Hew Strachan, "'Heroic' Warfare and the Problem of Mass Armies: France, 1871–1914," in *Heroism & the Changing Character of War,* ed. Sibylle Scheipers (Basingstoke, 2014), 47–63, esp. 49.

36. Samuel F. Scott, *The Response of the Royal Army to the French Revolution* (Oxford, 1978), 185–190; Forrest, *Conscripts and Deserters,* 23; Alan Forrest, *The Soldiers of*

the French Revolution (Durham, NC, 1990), 62–80; Roger Chickering, "A Tale of Two Tales: Grand Narratives of War in the Age of Revolution," in Chickering and Förster, *War in the Age of Revolution*, 7; and Ute Planert, "Innovation or Evolution? The French Wars in Military History," in Chickering and Förster, *War in the Age of Revolution*, 72.

37. Gooch, *Armies in Europe*, 2; Rapport, *Nationality and Citizenship*, 164–166, 215, 282–284.

38. Hippler, "Heroism and the Nation," 27; Planert, "Innovation or Evolution?," 73–76; Charles J. Esdaile, *The Wars of Napoleon* (London, 1995), 54–56; and Leighton S. James, "For the Fatherland? The Motivations of Austrian and Prussian Volunteers during the Revolutionary and Napoleonic Wars," in *War Volunteering in Modern Times: From the French Revolution to the Second World War*, ed. Christine G. Krüger and Sonja Levsen (Basingstoke, 2011), 40–58.

39. Peter Paret, "The Genesis of *On War*," in Clausewitz, *On War*, 9–18.

40. In September 1798 conscription was codified in France with the Loi Jourdan.

41. Planert, "Innovation or Evolution?," 78, 83; and Arthur Waldron, "Looking Backward: The People in Arms and the Transformation of War," in *The People in Arms: Military Myth and National Mobilization since the French Revolution*, ed. Daniel Moran and Arthur Waldron (Cambridge, 2003), 256–262.

42. R. D. Price, "The French Army and the Revolution of 1830," *European History Quarterly* 3 (1973): 243–267; Koller, "Recruitment Policies," 88.

43. J. H. Pryor, "The Oaths of the Leaders of the First Crusade to Emperor Alexius I Comnenus: Fealty, Homage—*pistis, douleia*," *Parergon* 2 (1984): 111–114; Jonathan Riley-Smith, *The First Crusaders, 1095–1131* (Cambridge, 1997); Christopher Tyerman, *Fighting for Christendom: Holy War and the Crusades* (Oxford, 2004); and Thomas S. Asbridge, *The First Crusade: A New History* (Oxford, 2004).

44. Tyerman, *Fighting for Christendom*, 27, 96.

45. Riley-Smith, *The First Crusaders*, 11.

46. Riley-Smith, *The First Crusaders*, 16–21.

47. Riley-Smith, *The First Crusaders*, 1, 8, 21–22, 83–85.

48. Tyerman, *Fighting for Christendom*, 134.

49. Asbridge, *The First Crusade*, 55.

50. Asbridge, *The First Crusade*, 61.

51. Asbridge, *The First Crusade*, 50–56; and Guy Perry, *John of Brienne: King of Jerusalem, Emperor of Constantinople, c. 1175–1237* (Cambridge, 2013), 21–23, 33.

52. Tilly, *Coercion*, 67.

53. Oswald Tesimond, *The Gunpowder Plot: The Narrative of Oswald Tesimond, Alias Greenway*, ed. Francis Edwards (London, 1973), 68.

54. Albert J. Loomie, *The Spanish Elizabethans: The English Exiles at the Court of Philip II* (London, 1963), 83, 177–178; and Mark Nicholls, "Fawkes, Guy," in *Oxford*

Dictionary of National Biography (Oxford, 2009), http://www.oxforddnb.com/index/101009230/Guy-Fawkes.

55. For more on questions of loyalty in the wake of the gunpowder plot, see Edward Vallance, *Revolutionary England and the National Covenant: State Oaths, Protestantism and the Political Nation, 1553–1682* (Woodbridge, UK, 2005), esp. 26.

56. John Childs, "The Williamite War, 1688–1691," in *A Military History of Ireland*, ed. Thomas Bartlett and Keith Jeffery (Cambridge, 1996), 188–210; Harman Murtagh, "Irish Soldiers Abroad, 1600–1800," in Bartlett and Jeffery, *A Military History of Ireland*, 294–314; Baron Godbert de Ginkel, "The Treaty of Limerick, 1691," The Corpus of Electronic Texts, http://www.ucc.ie/celt/online/E703001-010/text001.html; and James H. Murphy, "The Wild Geese," *Irish Review* 16 (1994): 23–28.

57. Thomas Bartlett and Keith Jeffery, "An Irish Military Tradition?," in Bartlett and Jeffery, *A Military History of Ireland*, 13. See also S. J. Connolly, *Divided Kingdom: Ireland 1630–1800* (Oxford, 2008), 286–290, 376; and Scott, "French Revolution," 14–24.

58. The tightening of restrictions on foreign enlistment did not develop in a linear manner. In 1730, when relations between Britain and France improved, the British government gave permission to the French to recruit up to 750 men in Ireland. See David Dickson, "Jacobitism in Eighteenth-Century Ireland: A Munster Perspective," *Éire-Ireland* 39, no. 3–4 (2004): 38–99.

59. Anno 9, George II, c. 30, in *The Statutes at Large of England and of Great Britain*, vol. 5, ed. John Raithby (London, 1811), 232; Murtagh, "Irish Soldiers Abroad," 310; Anno 29, George II, c. 17, in *The Statutes at Large of England and of Great Britain*, vol. 6, ed. John Raithby (London, 1811), 180–181.

60. Scott, "The French Revolution," 14–24; Bartlett and Jeffery, "An Irish Military Tradition," 11–12. For interesting parallels in practices of foreign enlistment elsewhere in the British Isles, see Stephen Conway, "Scots, Britons and Europeans: Scottish Military Service, c. 1739–1783," *Historical Research* 82, no. 215 (2009): 114–130.

61. See, for instance, Tarak Barkawi, "State and Armed Force in International Context," in *Mercenaries, Pirates, Bandits and Empire*, ed. Alejandro Colás and Bryan Mabee (London, 2010), 33–54; and Nir Arielli and Bruce Collins, "Introduction: Transnational Military Service since the Eighteenth Century," in Arielli and Collins, *Transnational Soldiers*, 1–10.

62. Marie Joseph Paul Yves Roch Gilbert Du Motier Lafayette, *Lafayette in the Age of the American Revolution: Selected Letters and Papers, 1776–1790*, vol. 1, ed. Stanley J. Idzerda (Ithaca, NY, 1977), xxiv–xxvi.

63. Catherine M. Prelinger, "Less Lucky than Lafayette: A Note on the French Applicants to Benjamin Franklin for Commissions in the American Army 1776–1785," *Proceedings of the Western Society for French History* 4 (1976): 263–270.

64. George H. Bushnell, *Kościuszko: A Short Biography of the Polish Patriot* (St. Andrews, Scotland, 1943), 5–9; and Peter J. Guthorn, "Kosciuszko as Military Cartographer and Engineer in America," *Imago Mundi* 29 (1977): 49–53.

65. Washington to Morris, 24 July 1778, in *The Writings of George Washington*, vol. 6, ed. Jared Sparks (Boston, 1834), 14–15.

66. Francis E. Zapatka, "Kościuszko in Early American Literature," *Polish American Studies* 47, no. 1 (1990): 55.

67. Guthorn, "Kosciuszko as Military Cartographer," 49–53; Bushnell, *Kościuszko*, 9–54.

68. "Memoir of 1779," in Lafayette, *Lafayette in the Age of the American Revolution*, 6–7.

69. "Agreement with Silas Deane," 7 December 1776, in Lafayette, *Lafayette in the Age of the American Revolution*, 17.

70. "Memoir of 1779," in Lafayette, *Lafayette in the Age of the American Revolution*, 11.

71. "Resolution of Congress," 31 July 1777, in Lafayette, *Lafayette in the Age of the American Revolution*, 88.

72. Cited in Eloise Ellery, *Brissot De Warville: A Study in the History of the French Revolution* (New York, 1970), 231.

73. Cited in Ellery, *Brissot De Warville*, 233.

74. Cited in Hippler, "Heroism and the Nation," 24.

75. Arthur Chuquet, *La Légion germanique* (Paris, 1904), 2–19; Simon Schama, *Patriots and Liberators: Revolution in the Netherlands 1780–1813* (London, 1977), 152; Rapport, *Nationality and Citizenship*, 158, 166; and Tozzi, *Nationalizing France's Army*, 89–103.

76. Brissot to Dumouriez, 28 November 1792, in *French Revolution Documents*, vol. 2, ed. John Hardman (Oxford, 1973), 391.

77. Charles-François Dumouriez, *Memoirs of General Dumourier* [*sic*], part 2, trans. John Fenwick (London, 1794), 29–31; and Schama, *Patriots and Liberators*, 153.

78. Owen Connelly, *The Wars of the French Revolution and Napoleon, 1792–1815* (London, 2005), 34–36; and Rapport, *Nationality and Citizenship*, 211–213, 274.

79. Karen Racine, *Francisco de Miranda: A Transatlantic Life in the Age of Revolution* (Wilmington, DE, 2003), xv.

80. Francisco De Miranda, *The Diary of Francisco De Miranda: Tour of the United States 1783–1784*, ed. William Spence Robertson (New York, 1928), 129.

81. Racine, *Francisco de Miranda*, 91.

82. Racine, *Francisco de Miranda*, 114.

83. Dumouriez, *Memoirs*, part 2, 18.

84. Racine, *Francisco de Miranda*, 120–140, 200–252.

85. A. Owen Aldridge, "John Oswald and the French Revolution," *The Eighteenth Century* 31, no. 2 (1990): 183; and Anna Plassart, "A Scottish Jacobin: John Oswald

on Commerce and Citizenship," *Journal of the History of Ideas* 71, no. 2 (2010): 265–268.

86. Plassart, "A Scottish Jacobin," 283–284.

87. Aldridge, "John Oswald," 186; Robert Roswell Palmer, *The Age of Democratic Revolution*, vol. 2, *The Struggle* (Princeton, NJ, 1964), 56; and Rapport, *Nationality and Citizenship*, 211.

88. Tozzi, *Nationalizing France's Army*, 156.

89. Cited in Rapport, *Nationality and Citizenship*, 278. See also Viktor Sautin, "The Polish Question: An Apple of Discord between Napoleon Bonaparte and Alexander I," *West Bohemian Historical Review* 2 (2011): 27–51; and Ruth Leiserowitz, "Polish Volunteers in the Napoleonic Wars," in *War Volunteering in Modern Times: From the French Revolution to the Second World War*, ed. Christine G. Krüger and Sonja Levsen (Basingstoke, 2011), 59–77.

90. Cited in Philippe R. Girard, *The Slaves Who Defeated Napoléon* (Tuscaloosa, AL, 2011), 206.

91. Cited in Bushnell, *Kościuszko*, 49. See also Jan Pachoński and Reuel K. Wilson, *Poland's Caribbean Tragedy: A Study of Polish Legions in the Haitian War of Independence 1802–1803* (Boulder, CO, 1986).

92. John G. Gallaher, *Napoleon's Irish Legion* (Carbondale, IL, 1993).

2. Attractive Conflicts

1. *Spain in Torment* (pamphlet), 1937, New York University Tamiment Library, Abraham Lincoln Brigade Archives collection (ALBA), VF, box 1, folder 18, "American Friends of Spanish Democracy."

2. Martin F. Shapiro, "Medical Aid Provided by American, Canadian and British Nationals to the Spanish Republic during the Civil War, 1936–1939," *International Journal of Health Services* 13, no. 3 (1983): 443–458; and Magí Crusells, "Refugee, a Pro-Republican Documentary from the Spanish Civil War," *International Journal of Iberian Studies* 19, no. 3 (2006): 231–239.

3. Fred Thomas, *To Tilt at Windmills* (East Lansing, MI, 1996), 5.

4. See, for instance, Michael Alpert, *A New International History of the Spanish Civil War* (Basingstoke, 2004), 127. For more on atrocities committed by supporters of the Spanish Republic, see Julio de la Cueva, "Religious Persecution, Anticlerical Tradition and Revolution: On Atrocities against the Clergy during the Spanish Civil War," *Journal of Contemporary History* 33, no. 3 (1998): 355–369. The number of those murdered on both sides of the conflict remains disputed. See, for instance, Julius Ruiz, "Defending the Republic: The García Atadell Brigade in Madrid, 1936," *Journal of Contemporary History* 42, no. 1 (2007): 97.

5. See, for instance, Derek J. Penslar, *Jews and the Military: A History* (Princeton, NJ, 2013), 201–251; and Charles Tripp, "Iraq and the War of 1948: Mirror of Iraq's Disorder," in *The War for Palestine: Rewriting the History of 1948,* ed. Eugene L. Rogan and Avi Shlaim (Cambridge, 2001), 131.

6. David C. Rapoport, "The Four Waves of Modern Terrorism," in *Attacking Terrorism: Elements of a Grand Strategy,* ed. Audrey Kurth Cronin and James M. Ludes (Washington, DC, 2004), 47.

7. Pierre Sautreuil, "Des paras français dans le Donbass," *Le Monde,* 26 August 2014, http://www.lemonde.fr/europe/article/2014/08/26/des-paras-francais-dans-le -donbass_4476646_3214.html; and Patrick Jackson, "Ukraine War Pulls in Foreign Fighters," BBC News, 1 September 2014, http://www.bbc.com/news/world -europe-28951324.

8. Mikulas Fabry, *Recognizing States: International Society and the Establishment of New States since 1776* (Oxford, 2010), 51–52; and Racine, *Francisco de Miranda,* 222.

9. Cited in Fabry, *Recognizing States,* 56.

10. Hansard, "Recognition of the Independence of South America—London Petition for," House of Commons debate, 15 June 1824, vol. 11 cc. 1344–1406.

11. Cited in Brown, *Adventuring through Spanish Colonies,* 113.

12. D. A. G. Waddell, "British Neutrality and Spanish-American Independence: The Problem of Foreign Enlistment," *Journal of Latin American Studies* 19, no. 1 (1987): 1–18.

13. Graciela Iglesias Rogers, *British Liberators in the Age of Napoleon: Volunteering under the Spanish Flag in the Peninsular War* (London, 2013), 161–162.

14. Brian Vale, *Cochrane in the Pacific: Fortune and Freedom in Spanish America* (London, 2008), esp. 26–140; James C. Carey, "Lord Cochrane: Critic of San Martín's Peruvian Campaign," *The Americas* 18, no. 4 (1962): 340–351; and William L. Neumann, "United States Aid to the Chilean Wars of Independence," *Hispanic American Historical Review* 27, no. 2 (1947): 218–219.

15. Cited in Brian Vale, *Independence or Death! British Sailors and Brazilian Independence, 1822–1825* (London, 1996), 18.

16. See, for instance, Matthew Brown, "Crusaders for Liberty or Vile Mercenaries? The Irish Legion in Columbia," *Irish Migration Studies in Latin America* 4, no. 2 (2006): 37–44; and Matthew Brown, "Soldier Heroes and the Columbian Wars of Independence," *Hispanic Research Journal* 7, no. 1 (2006): 41–56.

17. Cited in William St. Clair, *That Greece Might Still Be Free: The Philhellenes in the War of Independence* (London, 1972), 13.

18. For a critical analysis of early-nineteenth-century European admiration toward ancient Greece, see Elizabeth Roberts, *"Freedom, Faction, Fame and Blood": British Soldiers of Conscience in Greece, Spain and Finland* (Brighton, 2010), esp. 22–23. For

the North American variant, see Angelo Repousis, *Greek–American Relations from Monroe to Truman* (Kent, OH, 2013), esp. 10–26.

19. Lord Byron, *Childe Harold's Pilgrimage,* second canto, II (Chicago, 1900), 69.

20. Byron to John Murray, 7 February 1821, in *Life, Letters and Journals of Lord Byron,* ed. Thomas Moore (London, 1839), 693.

21. Davide Rodogno, *Against Massacre: Humanitarian Intervention in the Ottoman Empire 1815–1914* (Princeton, NJ, 2012), 65–67.

22. "London, Friday, June 1, 1821," *Times* [London], 1 June 1821, 3.

23. Gilles Pécout, "Philhellenism in Italy: Political Friendship and the Italian Volunteers in the Mediterranean in the Nineteenth Century," *Journal of Modern Italian Studies* 9, no. 4 (2004): 408; and J. L. Talmon, *Romanticism and Revolt: Europe 1815–1848* (London, 1967), 110–112.

24. Gordon, *History of the Greek Revolution,* 294, 316, 431; and St. Clair, *That Greece Might Still Be Free,* 32.

25. Lucy Riall, *Garibaldi: Invention of a Hero* (New Haven, CT, 2007), esp. 33–38; and Ronald Sarti, "Giuseppe Mazzini and Young Europe," in *Giuseppe Mazzini and the Globalisation of Democratic Nationalism 1830–1920,* ed. C. A. Bayly and Eugenio F. Biagini (Oxford, 2008), 280.

26. Riall, *Garibaldi,* 8, 39–45; and David McLean, "Garibaldi in Uruguay: A Reputation Reconsidered," *English Historical Review* 113, no. 451 (1998): 351–366.

27. "The Glorious Success of Garibaldi," *Huddersfield Chronicle and West Yorkshire Advertiser,* 9 June 1860, 4.

28. Cited in Don H. Doyle, *The Cause of All Nations: An International History of the American Civil War* (New York, 2014), 20.

29. Frank W. Alduino and David J. Coles, *Sons of Garibaldi in Blue and Gray: Italians in the American Civil War* (Youngstown, NY, 2007), 31–38.

30. "London, Thursday, Sept. 6, 1860," *Morning Post,* 6 September 1860, 4.

31. Cited in Riall, *Garibaldi,* 254. See also E. Boyer, "Les Volontaires français avec Garibaldi en 1860," *Revue d'histoire moderne et contemporaine* 7 (1960): 123–149.

32. Marcella Pellegrino Sutcliffe, "Negotiating the 'Garibaldi Moment' in Newcastle-upon-Tyne (1854–1861)," *Modern Italy* 15, no. 2 (2010): 129–144.

33. Charles A. Coulombe, *The Pope's Legion: The Multinational Fighting Force that Defended the Vatican* (New York, 2008), esp. 11.

34. The cause of liberty was also invoked, to a certain extent, during the civil wars in Portugal and Spain in the 1830s. For more on the participation of foreigners in these wars, see Grégoire Bron, "The Exiles of the Risorgimento: Italian Volunteers in the Portuguese Civil War (1832–34)," *Journal of Modern Italian Studies* 14, no. 4 (2009): 427–444; and Martin Robson, "'Strangers, Mercenaries, Heretics, Scoffers, Polluters': Volunteering for the British Auxiliary Legion in Spain,

1835," in *Transnational Soldiers: Foreign Military Enlistment in the Modern Era*, ed. Nir Arielli and Bruce Collins (Basingstoke, 2012), 181–201.

35. Eric R. Terzuolo, "The Garibaldini in the Balkans, 1875–1876," *International History Review* 4, no. 1 (1982): 111–126; and Pécout, "Philhellenism in Italy," 406.

36. Peter Whitewood, "Nationalities in a Class War: 'Foreign' Soldiers in the Red Army during the Russian Civil War," *Journal of Modern European History* 14, no. 3 (2016): 342–358.

37. Reinhard Nachtigal, "The Repatriation and Reception of Returning Prisoners of War, 1918–22," *Immigrants & Minorities* 26, no. 1–2 (2008): 157–184; and Ivan Volgyes, "Hungarian Prisoners of War in Russia 1916–1919," *Cahiers du monde russe et soviétique* 14, no. 1 (1973): 54–85.

38. Norman Davies, "The Missing Revolutionary War," *Soviet Studies* 27, no. 2 (1975): 180.

39. Kenneth Malcolm Murray, *Wings over Poland: The Story of the 7th (Kosciuszko) Squadron of the Polish Air Service, 1919, 1920, 1921* (New York, 1932), vii.

40. For more on Pilsudski's socialist past, see R. F. Leslie, ed., *The History of Poland since 1863* (Cambridge, 1983), 80–81.

41. Murray, *Wings over Poland,* ix; emphasis in the original.

42. Cited in James K. Hopkins, *Into the Heart of the Fire: The British in the Spanish Civil War* (Stanford, CA, 1998), 118.

43. Letter by Joseph Dallet, 19 March 1937, ALBA, Joseph Dallet collection, box 1, folder 11, correspondence, 1937.

44. Tom Wintringham, *English Captain* (London, 1939), 17.

45. Bill Alexander to Bill Moore, 15 [September] 1998, Sheffield City Archives, X274 Acc. 2009 / 118, box 15.

46. Cited in Hugo García, *The Truth about Spain! Mobilizing British Public Opinion 1936–1939* (Eastbourne, UK, 2010), 104.

47. García, *The Truth about Spain!,* 104–105.

48. Eoin O'Duffy, *Crusade in Spain* (Dublin, 1938), 12.

49. O'Duffy, *Crusade in Spain,* 11, 13.

50. George Orwell, *Homage to Catalonia* (London, 1971), 139.

51. Judith Keene, *Fighting for Franco: International Volunteers in Nationalist Spain during the Spanish Civil War* (London, 2007), esp. 291–292.

52. Steven O'Connor and Martin Gutmann, "Under a Foreign Flag: Integrating Foreign Units and Personnel in the British and German Armed Forces, 1940–1945," *Journal of Modern European History* 14, no. 3 (2016): 321–341.

53. Jean-Luc Leleu, "From the Nazi Party's Shock Troop to the 'European' Mass Army: The *Waffen-SS* Volunteers," in *War Volunteering in Modern Times: From the French Revolution to the Second World War,* ed. Christine G. Krüger and Sonja Levsen (Basingstoke, UK, 2011), 231–247.

54. Martin Gutmann, "Debunking the Myth of the Volunteers: Transnational Volunteering in the Nazi Waffen-SS Officer Corps during the Second World War," *Contemporary European History* 22, no. 4 (2013): 592.

55. David Motadel, *Islam and Nazi Germany's War* (Cambridge, MA, 2014), 2, 220–221; and O'Connor and Gutmann, "Under a Foreign Flag," 334.

56. Gutmann, "Debunking the Myth," 599; and Motadel, *Islam and Nazi Germany's War*, 3.

57. *Soldiers of the British Commonwealth! Soldiers of the United States of America!* (pamphlet) [1945], The National Archives of the United Kingdom (TNA), KV 2 / 2828, "The British Free Corps."

58. Riall, *Garibaldi*, 8; and Lucía Álvarez de Toledo, Introduction to *Bolivian Diary*, by Ernesto Che Guevara (London, 2004), xii–xiv.

59. Cited in Richard L. Harris, *Che Guevara: A Biography* (Santa Barbara, CA, 2011), 51.

60. Ernesto Guevara, "Episodes of the Revolutionary War," in *Che Guevara Reader: Writings on Politics & Revolution,* ed. David Deutschmann (Melbourne, 2003), 20.

61. Cited in Jon Lee Anderson, *Che Guevara: A Revolutionary Life* (New York, 2010), 184.

62. "Message to the Tricontinental," [April 1967], in Deutschmann, *Che Guevara Reader,* 354.

63. "Speech to Medical Students and Health Workers," 20 August 1960, in Deutschmann, *Che Guevara Reader,* 115.

64. Richard Gott, Introduction to *The African Dream: The Diaries of the Revolutionary War in the Congo,* by Ernesto "Che" Guevara (London, 2000), xl.

65. Guevara, *The African Dream,* 27, 57–67; Harris, *Che Guevara,* 138.

66. Guevara, *The African Dream,* 7.

67. Guevara, *The African Dream,* 1. The Cuban force that accompanied Guevara does not fit my definition of foreign volunteers because its members were sent to Congo by the government in Havana. This was part of a broader Cuban strategy in Africa. Between the early 1960s and the late 1980s, Castro's regime sent both military and civilian assistance to the Algerian republic, the insurgents in Guinea-Bissau and the People's Movement for the Liberation of Angola. Piero Gleijeses, *Conflicting Missions: Havana, Washington and Africa, 1959–1976* (Chapel Hill, NC, 2002); Piero Gleijeses, *Visions of Freedom: Havana, Washington, Pretoria, and the Struggle for Southern Africa, 1976–1991* (Chapel Hill, NC, 2013).

68. Cited in Fidel Castro, "A Necessary Introduction," in *Bolivian Diary,* by Ernesto Che Guevara (London, 2004), xxxiv.

69. Jim Morris, Foreword to *Save the Last Bullet for Yourself: A Soldier of Fortune in the Balkans and Somalia,* by Rob Krott (Drexel Hill, PA, 2008), xi; and Farrelly, "Transnational Flows of Military Talent," 152. Jean-Philippe Courrèges was

killed in action. The cross over his grave bears the epitaph: "Jean-Philippe Memory of Karen People Forever."

70. Andrea Wolf was killed in 1998. "PKK Member Andrea Wolf Commemorated," ANF News, 17 June 2015, http://www.anfenglish.com/kurdistan/pkk-member -andrea-wolf-commemorated.

71. Samuel P. Huntington, "The Clash of Civilizations?," *Foreign Affairs* 72, no. 3 (1993): 22.

72. There are exceptions. In 2011, Omar Shafik Hammami, also known as Abu Man-soor al-Amriki, an American volunteer in the ranks of the Somali al-Shabaab, issued a recruitment video in which he says, "This is a war of civilizations, it's not a war of individuals." See "'Lessons Learned'—By Mujahid Commander: Abu Mansoor al-Amriki Part 1," YouTube, posted 7 October 2011, https://www .youtube.com/watch?v=z52zYdo_beA.

73. Asaf Maliach and Shaul Shay, *From Kabul to Jerusalem* (Tel Aviv, 2009), 22–34 [Hebrew].

74. Maliach and Shay, *From Kabul to Jerusalem*, 35–79; Thomas Hegghammer, "The Rise of Muslim Foreign Fighters," *International Security* 35, no. 3 (2011): 85–88; and Peter L. Bergen, *The Osama bin Laden I Know: An Oral History of Al Qaeda's Leader* (New York, 2014), 39.

75. See, for instance, the excerpt from *Jihad*, 1, 28 December 1984, cited in Bergen, *The Osama bin Laden I Know*, 33.

76. Abdullah Azzam, *Join the Caravan* (London, 2001), 19.

77. Azzam, *Join the Caravan*, 50–51.

78. *Jihad*, 3, 22 March 1985, cited in Bergen, *The Osama bin Laden I Know*, 42.

79. See Maliach and Shay, *From Kabul to Jerusalem*, 65, 89; and Hegghammer, "The Rise of Muslim Foreign Fighters," 74–75.

80. See Maliach and Shay, *From Kabul to Jerusalem*, 43–44; Hegghammer, "The Rise of Muslim Foreign Fighters," 61; Bergen, *The Osama bin Laden I Know*, 41; and Peter L. Bergen, *Holy War, Inc.* (London, 2002), 58. The term "Afghan Arabs" or, as it some-times appears, "Arab Afghans," is misleading because volunteers also came from non-Arab countries.

81. John Glen, dir., *The Living Daylights* (Eon Productions, 1987); Peter MacDonald, dir., *Rambo III* (Taito, 1988).

82. Samuel Huntington, *The Clash of Civilizations and the Remaking of World Order* (New York, 2003 [1996]), 271. Depicting the war in former Yugoslavia as a "neat" confrontation between three homogenous groups oversimplifies the conflict. Alignments were often much more fragmented. In the northwestern Bosnian region of Krajina, for instance, separatist Bosnian Muslims cooperated with Serb forces and fought against Bosnian government troops.

83. Ali M. Koknar, "The Kontraktniki: Russian Mercenaries at War in the Balkans," website of the Bosnian Institute, 14 July 2003, www.bosnia.org.uk/news/news _body.cfm?newsid=1766; Aziz Tafro, *Ruski i Grčki Plaćenici u Ratu u Bosni i Her-cegovini* (Sarajevo, 2014), esp. 32; and Helena Smith, "Greece Faces Shame of Role in Serb Massacre," *Guardian*, 5 January 2003, http://www.theguardian.com /world/2003/jan/05/balkans.warcrimes.

84. Marc Bennetts, "Eduard Limonov Interview: Political Rebel and Vladimir's Putin Worst Nightmare," *Guardian*, 12 December 2010, http://www.theguardian .com/world/2010/dec/12/eduard-limonov-interview-putin-nightmare.

85. Nir Arielli, "In Search of Meaning: Foreign Volunteers in the Croatian Armed Forces, 1991–1995," *Contemporary European History* 21, no. 1 (2012): 1–17.

86. Darryl Li, "Jihad in a World of Sovereigns: Law, Violence, and Islam in the Bosnia Crisis," *Law & Social Inquiry* 41, no. 2 (2016): 371–401.

87. See, for instance, Evan Kohlmann, *Al-Qaida's Jihad in Europe: The Afghan-Bosnian Network* (Oxford, 2004).

88. Bergen, *The Osama bin Laden I Know*, 74, 164; and Lawrence Wright, *The Looming Tower: Al-Qaeda's Road to 9 / 11* (London, 2007), 302.

89. Maliach and Shay, *From Kabul to Jerusalem*, 194–195.

90. CounterTerrorismUnit, "British Jihadist in Bosnia: Why I Want to Fight Jihad," YouTube, posted on 5 June 2011, https://www.youtube.com/watch?v=Y629 EEgIJ34.

91. "The Spy Who Came in from Al-Qaeda," BBC4, 3 March 2015, http://www.bbc .co.uk/programmes/b05mrj7f.

92. Cited in Bergen, *The Osama bin Laden I Know*, 164–165.

93. Thomas Hegghammer, "Should I Stay or Should I Go? Explaining Variation in Western Jihadists' Choice between Domestic and Foreign Fighting," *American Political Science Review* 107, no. 1 (2013): 1–15.

94. Cited in Peter R. Neumann, *Joining al-Qaeda: Jihadist Recruitment in Europe* (Abingdon, UK, 2008), 50.

95. Hegghammer, "Should I Stay or Should I Go?," 10; and Wright, *The Looming Tower*, 301–307.

96. Hegghammer, "The Rise of Muslim Foreign Fighters," 61. See also Jane Ferguson, "Violent Extremists Calling Fighters to Somalia," CNN, 27 April 2010, http://edition.cnn.com/2010/WORLD/africa/04/27/somalia.al.shabaab/index .html. The conflict in Iraq received greater media coverage and was more easily accessible for foreign volunteers in comparison to Somalia and Afghanistan.

97. Cited in Bergen, *The Osama bin Laden I Know*, 353.

98. Azzam, *Join the Caravan*, 33. In 2001 the future leader of al-Qaeda, Ayman al-Zawahiri, explained why he had been attracted to Afghanistan in the early 1980s: "A jihadist movement needs an arena that would act like an incubator

where its seeds would grow and where it can acquire experience in combat, politics, and organizational matters." He had sought such a "secure base" in his native Egypt, but repression by the regime made this impossible. Laura Mansfield, ed., *His Own Words: A Translation of the Writings of Dr. Ayman Al Zawahiri* (Old Tappan, NJ, 2006), 35.

99. Mohammed M. Hafez, *Suicide Bombers in Iraq* (Washington, DC, 2007), 72–78; Bergen, *The Osama bin Laden I Know*, 350–367; and Wright, *The Looming Tower*, 301.

100. Abu Bakr Al-Baghdadi, speech, "March Forth Whether Light or Heavy," released on 14 May 2015, *MEMRI* website, http://www.memrijttm.org/in-new-audio -speech-islamic-state-isis-leader-al-baghdadi-issues-call-to-arms-to-all-muslims .html.

101. Some of the foreign fighters in Syria have been referred to as al-Muhajiroun (the emigrants).

102. Jennifer Percy, "Meet the American Vigilantes Who Are Fighting ISIS," *New York Times Magazine*, 30 September 2015, http://www.nytimes.com/2015/10/04/magazine /meet-the-american-vigilantes-who-are-fighting-isis.html?_r=1; and Mark Tran, "Briton Who Spent Months Fighting ISIS Wanted to 'Shine a Light' on the Conflict," *Guardian*, 11 June 2015, http://www.theguardian.com/world/2015 /jun/11/macer-gifford-briton-fight-isis-shine-light-conflict.

103. Nick Logan, "John Gallagher: Here's What We Know about the Canadian Killed While Fighting ISIS in Syria," *Global News*, 5 November 2015, http://globalnews .ca/news/2321304/john-gallagher-heres-what-we-know-about-the-canadian -killed-while-fighting-isis-in-syria/.

104. John Gallagher, "Why the War in Kurdistan Matters," Facebook post, 6 May 2015, https://www.facebook.com/permalink.php?story_fbid=10152939717688845&id =642603844.

3. A Search for Meaning

1. Norman Cigar, "Overview and Analysis," in *Al-Qa'ida's Doctrine for Insurgency* (Washington, DC, 2009), 6–7; and Elena Mastors and Alyssa Deffenbaugh, *The Lesser Jihad: Recruits and the Al-Qaida Network* (Lanham, MD, 2007), 119n33.

2. Jarret M. Brachman, *Global Jihadism: Theory and Practice* (London, 2009), 3; and Cigar, "Overview and Analysis," 6.

3. Maliach and Shay, *From Kabul to Jerusalem*, 44.

4. Penslar, *Jews and the Military*, 241–246.

5. Michael Jackson, *Fallen Sparrows: The International Brigades in the Spanish Civil War* (Philadelphia, 1994), 40.

6. See John MacPhee, *The Silent Cry: One Man's Fight for Croatia in the Bosnian War* (Manchester, 2000), esp. 173–181; and Krott, *Save the Last Bullet*, 197.

7. Penslar, *Jews and the Military*, 234 and 310n25. For a critical appraisal of Dunkelman's conduct during the war of 1948, see Dan Freeman-Maloy, "Mahal and the Dispossession of the Palestinians," *Journal of Palestine Studies* 40, no. 2 (2011): 43–61.

8. See, for instance, Jsrael Centner, *From Madrid to Berlin* (Tel Aviv, 1966), esp. 33–35, 53–55, 89–90 [Hebrew].

9. Sandes, *The Autobiography of a Woman Soldier*, 16.

10. Joseph Heller's interview with David Karon, Kfar Menachem, Israel, 5 July 1965, Oral History Database of the Hebrew University of Jerusalem, interview no. 38 / 1 [Hebrew].

11. Iglesias Rogers, *British Liberators*, 25.

12. See, for instance, Daniel Kahneman, *Thinking, Fast and Slow* (London, 2011).

13. Peter Kemp, *The Thorns of Memory* (London, 1990), 7.

14. "Britons Lured to Red Front," *Daily Mail*, 18 February 1937, 13. See also Richard Baxell, *Unlikely Warriors: The British in the Spanish Civil War and the Struggle against Fascism* (London, 2012), 56–57.

15. Herbert Greene, *Secret Agent in Spain* (London, 1938), 284.

16. J. Edgar Hoover, *Masters of Deceit: The Story of Communism in America and How to Fight It* (New York, 1958), 93.

17. For a useful summary of the historiographic dispute, see George Esenwein, "Freedom Fighters or Comintern Soldiers? Writing about the 'Good Fight' during the Spanish Civil War," *Civil Wars* 12, no. 1–2 (2010): 156–166.

18. R. Dan Richardson, *Comintern Army: The International Brigades and the Spanish Civil War* (Lexington, KY, 1982), 15.

19. Jackson, *Fallen Sparrows*, 40.

20. See, for instance, Ray Sanchez and Greg Botelho, "U.S. Investigates How Americans Are Lured to Syria," CNN, 29 August 2014, http://edition.cnn.com/2014/08/28/us/syria-americans-killed/.

21. See, for instance, Harriet Sherwood, Sandra Laville, Kim Willsher, Ben Knight, Maddy French, and Lauren Gambino, "Schoolgirl Jihadis: The Female Islamists Leaving Home to Join Isis Fighters," *Guardian*, 29 September 2014, http://www.theguardian.com/world/2014/sep/29/schoolgirl-jihadis-female-islamists-leaving-home-join-isis-iraq-syria.

22. Malet, *Foreign Fighters*, 4.

23. Malet, *Foreign Fighters*, 3, 5, 166.

24. Thomas, *To Tilt at Windmills*, 5–6.

25. "Howard Cossman," Volunteers from Overseas in the Israel Defense Forces, World Machal Website, http://www.machal.org.il/index.php?option=com_content&view=article&id=228&Itemid=316&lang=he.

26. CounterTerrorismUnit, "British Jihadist in Bosnia."

27. Thomas Hegghammer, "Saudis in Iraq: Patterns of Radicalization and Recruitment," *Cultures & Conflicts*, 12 June 2008, 8, http://conflits.revues.org/10042.

28. "Frontline Fighting: The Brits Battling Isis," Channel 4, 16 September 2015, http://www.channel4.com/programmes/frontline-fighting-the-brits-battling-isis.

29. Robin D. G. Kelly, "This Ain't Ethiopia but It'll Do," in *African Americans in the Spanish Civil War*, ed. Danny Duncan Collum (New York, 1992), 16–18; and James Yates, *Mississippi to Madrid: Memoir of a Black American in the Abraham Lincoln Brigade* (Seattle, 1989), 89.

30. Cited in Kelly, "This Ain't Ethiopia but It'll Do," 6.

31. Yates, *Mississippi to Madrid*, 91.

32. *Salaria Kea: A Negro Nurse in Republican Spain* (pamphlet) [1938], Abraham Lincoln Brigades Archives website, http://www.alba-valb.org/resources/robeson-primary-resources/salaria-kea-a-negro-nurse-in-republican-spain.

33. Cited in Hywel Francis, "Welsh Miners and the Spanish Civil War," *Journal of Contemporary History* 5, no. 3 (1970): 188.

34. Hyman Katz to "Ma," 25 November 1937, in *Madrid 1937: Letters of the Abraham Lincoln Brigade from the Spanish Civil War*, ed. Cary Nelson and Jefferson Hendricks (New York, 1996), 32. For more on how the Spanish Civil War was interpreted by various Jewish communities, see Penslar, *Jews and the Military*, 200–207; and Gerben Zaagsma, *Jewish Volunteers, the International Brigades and the Spanish Civil War* (London, 2017).

35. Author's interview with Steve Gaunt, Vinkovci, Croatia, 14 March 2010.

36. Steve Gaunt, *War and Beer* (Coventry, UK, 2010), 4.

37. Gaunt, *War and Beer*, 4.

38. Simon Hutt, *Paint: A Boy Soldier's Journey* (Coventry, UK, 2010), 82, 88.

39. Mary Habeck, *Knowing the Enemy: Jihadist Ideology and the War on Terror* (New Haven, CT, 2006), 7.

40. Vikram Dodd, "Two British Men Admit to Linking Up with Extremist Groups in Syria," *Guardian*, 8 July 2014, http://www.theguardian.com/world/2014/jul/08/two-british-men-admit-linking-extremist-group-syria.

41. Lydia Wilson, "What I Discovered from Interviewing Imprisoned ISIS Fighters," *Nation*, 21 October 2015, http://www.thenation.com/article/what-i-discovered-from-interviewing-isis-prisoners/.

42. Kahneman, *Thinking, Fast and Slow*, 87.

43. "Former British Soldier Plans to Fight IS in Syria," BBC Radio 4, 10 March 2015. A short excerpt from the interview is available online: http://www.bbc.co.uk/programmes/po2lnjqg.

44. Francis, "Welsh Miners," esp. 182; and Fraser Raeburn, "'Fae nae hair Te grey hair they answered the call': International Brigade Volunteers from the West

Central Belt of Scotland in the Spanish Civil War, 1936–9," *Journal of Scottish Historical Studies* 35, no. 1 (2015): 92–114. Clusters of volunteers who come from the same cities or towns and know each other prior to joining a foreign conflict have also been observed in other instances. See Marcella Pellegrino Sutcliffe, "British Red Shirts: A History of the Garibaldi Volunteers (1860)," in *Transnational Soldiers: Foreign Military Enlistment in the Modern Era*, ed. Nir Arielli and Bruce Collins (Basingstoke, 2012), 210–211; and Joseph Felter and Brian Fishman, *Al-Q'aida's Foreign Fighters in Iraq: A First Look at the Sinjar Records* (West Point, NY, 2007), 10–15.

45. Cited in Iglesias Rogers, *British Liberators*, 27.

46. Cited in Benita Eisler, *Byron: Child of Passion, Fool of Fame* (London, 1999), 721.

47. Cited in Roman Robert Koropeckyj, *Adam Mickiewicz: The Life of a Romantic* (Ithaca, NY, 2008), 443, 445.

48. F. M. Leventhal, *The Last Dissenter: H. N. Brailsford and His World* (Oxford, 1985), 28.

49. Gustav Regler, *The Owl of Minerva* (London, 1959), 266.

50. Lisa A. Kirschenbaum, *International Communism and the Spanish Civil War* (New York, 2015), 135.

51. Nir Arielli, "Induced to Volunteer: The Predicament of Jewish Communists in Palestine and the Spanish Civil War," *Journal of Contemporary History* 46, no. 4 (2011): 854–870.

52. See, for instance, Leah Trachtman-Palchan, *Between Tel Aviv and Moscow: A Life of Dissent and Exile in Mandate Palestine and the Soviet Union* (London, 2015).

53. Remarks by Arie Lev in "Why They Volunteered," *Proceedings of the Global Conference of the Jewish Fighters in the International Brigades in Spain* (Tel Aviv, 1972), 16 [Hebrew], Lavon Institute Archive, Tel Aviv, IV / 104 / 1256 / 1. The threat of deportation also hung over the heads of communist Polish-Jewish exiles who lived in Belgium. A large number of them decided to join the International Brigades rather than face lengthy imprisonment back in their country of birth. See Zaagsma, *Jewish Volunteers*, 20 and 23.

54. Cited in Francis, "Welsh Miners," 186.

55. Peter N. Carroll, *The Odyssey of the Abraham Lincoln Brigade: Americans in the Spanish Civil War* (Stanford, CA, 1994), 15–17; Michael Petrou, *Renegades: Canadians in the Spanish Civil War* (Vancouver, 2008), 12–25; Baxell, *Unlikely Warriors*, 13–25; and Raeburn, "Fae nae hair," 97–102.

56. Yates, *Mississippi to Madrid*, 41–89.

57. Jessica Stern and J. M. Berger, "ISIS and the Foreign Fighter Phenomenon," *Atlantic,* 8 March 2015, http://www.theatlantic.com/international/archive/2015/03/isis-and-the-foreign-fighter-problem/387166/.

58. Viktor E. Frankl, "Basic Concepts of Logotherapy," in *Man's Search for Meaning* (London, 1970), 105.

59. Frankl, "Basic Concepts," 106.

60. Frankl, "Basic Concepts," 115.

61. Frankl, "Basic Concepts," 136.

62. See, for instance, Henri Tajfel and John Turner, "An Integrative Theory of Intergroup Conflict," in *The Social Psychology of Intergroup Relations,* ed. W. G. Austin and S. Worchel (Monterey, CA, 1979), 40.

63. Pierre Bourdieu, *Pascalian Meditations* (Cambridge, 2000), 237.

64. Bourdieu, *Pascalian Meditations,* 240; emphasis in the original.

65. Bourdieu, *Pascalian Meditations,* 240–242.

66. Bourdieu, *Pascalian Meditations,* 240–241.

67. Cited in Koropeckyj, *Adam Mickiewicz,* 440.

68. Cited in Koropeckyj, *Adam Mickiewicz,* 453.

69. Regler, *The Owl of Minerva,* 41.

70. Regler, *The Owl of Minerva,* 271.

71. Regler, *The Owl of Minerva,* 272.

72. Harold Livingston, *No Trophy, No Sword: An American Volunteer in the Israeli Air Force during the 1948 War of Independence* (Chicago, 1994), 9.

73. Livingston, *No Trophy, No Sword,* 15.

74. Livingston, *No Trophy, No Sword,* 8.

75. Livingston, *No Trophy, No Sword,* 18.

76. Livingston, *No Trophy, No Sword,* 199.

77. "The Spy Who Came in from Al-Qaeda."

78. Author's correspondence with Roland Bartetzko, July 2015. See also Bartetzko's interview that features in the documentary film *Soldier-Ushtari* (2014), directed by Mrika Krasniqi.

79. Author's correspondence with Roland Bartetzko, July 2015.

80. See, for instance, the comments made by the British volunteer Jim in "Frontline Fighting."

81. Author's correspondence with Valerie Carder, December 2015.

82. John Gallagher's Facebook page, 4 August 2015, https://www.facebook.com /profile.php?id=642603844&fref=nf.

83. Author's correspondence with Valerie Carder, December 2015.

84. Frankl, "Basic Concepts of Logotherapy," 106.

85. Richard Baxell, *British Volunteers in the Spanish Civil War* (London, 2004), 25; and Angela Jackson, *"For Us It Was Heaven": The Passion, Grief and Fortitude of Patience Darton* (Brighton, 2012), 208n6.

86. "Machal Comes to Israel: A Study on the Adaptation of Overseas Volunteers to Israel and Its Inhabitants," 1949 [Hebrew]. The survey is available through the website of the Israel Democracy Institute, http://www.idi.org.il/.

87. Arielli, "In Search of Meaning," 2, 15. The approach toward wartime gender roles in the Balkans appear to have changed significantly since the Second World War.

In the 1940s, some 100,000 local women fought as partisans. Juraj Dobrila, "Partizanke: Their Dangerous Legacy in the Post-Yugoslav Space," Dangerous Women Project, 1 April 2016, http://dangerouswomenproject.org/2016/04/01/partizanke-dangerous-legacy/. In contrast, the wars of the 1990s were gendered to such an extent that the International Criminal Tribunal for the Former Yugoslavia classified rape as a crime against humanity and convicted people of using rape as a weapon of war.

88. James Hamilton Browne, "Voyage from Leghorn to Cephalonia with Lord Byron," *Blackwood's Edinburgh Magazine* 35 (1834): 64.

89. Peter Cochran, "Introduction: The Bisexual Byron," in *Byron and Women [and Men]*, ed. Peter Cochran (Newcastle upon Tyne, UK, 2010), xxviii.

90. Leventhal, *The Last Dissenter*, 28.

91. Henry Noel Brailsford, *The Broom of the War-God* (London, 1898), 155.

92. Kemp, *The Thorns of Memory*, 4.

93. See, for instance, Giuseppe Garibaldi, *A Toast to Rebellion* (London, 1936), esp. xi–xiii; and Anthony Loyd, *My War Gone By, I Miss It So* (London, 2002), esp. 57–58.

94. Interview featured in the film *Above and Beyond* (2015), directed by Roberta Grossman.

95. Guevara, *The African Dream*, 40.

96. 8 August 1967, in Ernesto Che Guevara, *Bolivian Diary* (London, 2004), 149.

97. Tamar Horowitz, Joseph Hodara, and Meir Cialic, "Volunteers for Israel during the Six Day War: Their Motives and Work Careers," *Dispersion and Unity* 13–14 (1971): 85.

98. See, for instance, Jeff Jones, ed., *Brigadista: Harvest and War in Nicaragua* (New York, 1986). Out of fifty-seven volunteers who contributed to this book, approximately 60 percent were women.

99. Cited in Jackson, *"For Us It Was Heaven,"* 22.

100. Author's interview with Lily Myerson, Nordea, Israel, 23 December 2010.

101. Author's interview with Anita Koifman, Tel Ganim, Israel, 22 December 2010.

102. Cited in Bergen, *The Osama bin Laden I Know*, 43.

103. Cited in Katherine Brown, "How IS Message Lures Western Women," BBC News, 8 April 2015, http://www.bbc.com/news/world-middle-east-32208217.

104. Cited in Benedetta Argentieri, "From Brides to Battlefield, Women Taking Up New Roles with Islamic State," Reuters, 1 April 2015, http://www.reuters.com/article/us-mideast-crisis-women-idUSKBN0MT00320150402. See also Joanna Witt, "Why Do Young Women Want to Join Islamic State?," *Guardian*, 27 July 2015, http://www.theguardian.com/membership/2015/jul/27/guardian-live-why-do-young-women-want-to-join-islamic-state; and Jayne Huckerby, "Why Women Join ISIS," *Time*, 7 December 2015, http://time.com/4138377/women-in-isis/.

105. Jasper Ridley, *Garibaldi* (London, 1974), 95, 302, 311, and 323.

106. Sandes, *The Autobiography of a Woman Soldier*, 9.

107. Flora Sandes, *An English Woman-Sergeant in the Serbian Army* (London, 1916), 23; and Janet Lee, "A Nurse and a Soldier: Gender, Class and National Identity in the First World War Adventures of Grace McDougall and Flora Sandes," *Women's History Review* 15, no. 1 (2006): 85.

108. Sandes, *The Autobiography of a Woman Soldier*, 13.

109. Sandes, *The Autobiography of a Woman Soldier*, 12–13.

110. Sandes, *The Autobiography of a Woman Soldier*, 17.

111. Yvonne Scholten, "Fanny, Queen of the Machine Gun," *Volunteer*, 4 December 2011, http://www.albavolunteer.org/2011/12/queen-of-the-machine-gun-fanny-schoonheyt-dutch-miliciana/.

112. "English Sculptress Killed in Spain," *Daily Worker*, 4 September 1936, reproduced in James Pettifer, *Cockburn in Spain: Despatches from the Spanish Civil War* (London, 1986), 63.

113. Cited in Tom Buchanan, *The Impact of the Spanish Civil War on Britain: War, Loss and Memory* (Brighton, UK, 2007), 78.

114. Centner, *From Madrid to Berlin*, 57–58.

115. Larry Hannant, "'My God, Are They Sending Women?' Three Canadian Women in the Spanish Civil War," *Journal of the Canadian Historical Association* 15, no. 1 (2004): 165; and Shirley Mangini, *Memories of Resistance: Women's Voices from the Spanish Civil War* (New Haven, CT, 1995), 80–81. For more on debates regarding the incorporation of women into military organizations, see Joshua Goldstein, *War and Gender: How Gender Shapes the War System and Vice Versa* (Cambridge, MA, 2011).

116. Zaagsma, *Jewish Volunteers*, 19.

117. Hannant, "'My God, Are They Sending Women?,'" 155.

118. See, for instance, "How Kurdish Woman Soldiers Are Confronting ISIS on the Front Lines," *PBS Newshour*, 3 May 2015, http://www.pbs.org/newshour/bb/kurdish-women-soldiers-confronting-fears-isis/.

119. Elhanan Miller, "IS Rape and Torture of Yezidi Women Pushed Me to Fight with the Kurds, Says Gill Rosenberg," *Times of Israel*, 17 July 2015, http://www.timesofisrael.com/is-rape-and-torture-of-yazidi-women-pushed-me-to-fight-with-kurds-says-gill-rosenberg/.

4. Thoughts of Home

1. Robin W. Winks, *The Civil War Years: Canada and the United States* (Montreal, 1998), 184–185.

2. Jason Gurney, *Crusade in Spain* (London, 1974), 66.

3. John M. Venhaus, "Why Youth Join al-Qaeda," United States Institute of Peace, *Special Report* 236 (2010): 1–20.

4. Percy, "Meet the American Vigilantes."

5. Rodogno, *Against Massacre*, 72–78.

6. Rodogno, *Against Massacre*, 126–127; and Leonidas F. Callivretakis, "Les Garibaldiens a l'insurrection de 1866 en Crète: Le jeu des chiffres," in *Indipendenza e unità nazionale in Italia ed in Grecia: Convegno di studio* (Florence, 1987), 163–177.

7. Pécout, "The International Armed Volunteers," 414.

8. David MacKenzie, "Panslavism in Practice: Cherniaev in Serbia (1876)," *Journal of Modern History* 36, no. 3 (1964): 279–297; and Walter G. Moss, *Russia in the Age of Alexander II, Tolstoy and Dostoevsky* (London, 2002), 180–181.

9. Cited in Roy Macnab, *The French Colonel: Villebois-Mareuil and the Boers 1899–1900* (Cape Town, 1975), 16.

10. Captain Alfred Dreyfus, a French artillery officer, was arrested in 1894 and falsely accused of passing military secrets to the Germans. His trial, conviction, and the subsequent public outcry sharply divided French society.

11. Macnab, *The French Colonel*, 51–70; and Fransjohan Pretorius, "Welcome but Not That Welcome: The Relations between Foreign Volunteers and the Boers in the Anglo-Boer War," in *War Volunteering in Modern Times: From the French Revolution to the Second World War*, ed. Christine G. Krüger and Sonja Levsen (Basingstoke, 2011), 124.

12. Georges Villebois-Mareuil, *War Notes: The Diary of Colonel de Villebois-Mareuil*, trans. Frederic Lees (London, 1901), 82.

13. E. M. de Vogüé, Introduction to Villebois-Mareuil, *War Notes*, xvii.

14. Villebois-Mareuil, *War Notes*, 282–283.

15. Charlotte Haldane, *Truth Will Out* (London, 1949), 101.

16. Cited in Baxell, *British Volunteers in the Spanish Civil War*, 26.

17. Geoffrey Cox, *The Red Army Moves* (London, 1941), 105.

18. Cox, *The Red Army Moves*, 105.

19. Cited in Martina Sprague, *Swedish Volunteers in the Russo-Finnish Winter War, 1939–1940* (Jefferson, NC, 2010), 69.

20. Gordon Levett, *Flying under Two Flags: An RAF Pilot in Israel's War of Independence* (London, 1994), 123.

21. Levett, *Flying under Two Flags*, 124–125.

22. Joel Greenberg, "'Fun Stuff' in 1948: British Gentile in Israeli Air Force," *New York Times*, 10 May 1998, http://www.nytimes.com/1998/05/10/world/fun-stuff-in-48-british-gentile-in-israel-air-force.html.

23. Author's correspondence with Ivan Farina, September 2010.

24. Héctor Perla, "Heirs of Sandino: The Nicaraguan Revolution and the U.S.-Nicaragua Solidarity Movement," *Latin American Perspective* 36, no. 6 (2009): 80–100.

25. Cited in Jones, *Brigadista*, 36.

26. See, for instance, Paul Hockenos, *Homeland Calling: Exile Patriotism & the Balkan Wars* (Ithaca, NY, 2003), 8; and Dina Al Raffie, "Social Identity Theory for Investigating Islamic Extremism in the Diaspora," *Journal of Strategic Security* 6, no. 4 (2013): 68.

27. Jolle Demmers, "New Wars and Diasporas: Suggestions for Research and Policy," *Journal of Peace, Conflict and Development* 11 (2007): 9.

28. Demmers, "New Wars and Diasporas," 8; Al Raffie, "Social Identity Theory," 74.

29. "Brief no. 15: Gun Running" and "Brief no. 16: Fund Raising," October 1977, TNA, CJ 4 / 1839.

30. St. Clair, *That Greece Might Still Be Free*, 25.

31. Thomas Keightley, *History of the War of Independence in Greece*, vol. 2 (Edinburgh, 1830), 48–50; St. Clair, *That Greece Might Still Be Free*, 26–34; and Andrew Mac-Gregor, *A Military History of Modern Egypt* (Westport, CT, 2006), 88.

32. Roberts, "Freedom, Faction, Fame and Blood," 19–21, 28.

33. Stephen Applebaum, "Interview: Vidal Sassoon," *Jewish Chronicle Online*, 12 May 2011, http://www.thejc.com/arts/arts-interviews/48769/interview-vidal-sassoon.

34. Vidal Sassoon, *Vidal: The Autobiography* (London, 2011), 59

35. Sassoon, *Vidal*, 59.

36. Livingston, *No Trophy, No Sword*, 18.

37. Livingston, *No Trophy, No Sword*, 18.

38. Razmik Panossian, "The Armenians: Conflicting Identities and the Politics of Division," in *Nations Abroad: Diaspora Politics and International Relations in the Former Soviet Union*, ed. Charles King and Neil J. Melvin (Boulder, CO, 1998), 79–102; Khachig Tölölyan, "The Armenian Diaspora and the Karabagh Conflict since 1988," in *Diasporas in Conflict: Peace-Makers or Peace-Wreckers?*, ed. Hazel Smith and Paul Stares (Tokyo, 2007), 106–128; and Maria Koinova, "Diasporas and Secessionist Conflicts: The Mobilization of the Armenian, Albanian and Chechen Diasporas," *Ethnic and Racial Studies* 34, no. 2 (2011): 333–356.

39. Hockenos, *Homeland Calling*, 3, 36.

40. Hockenos, *Homeland Calling*, 17–49. When he passed away, Šušak was under investigation by the International Criminal Tribunal for the Former Yugoslavia for his role in the war in Bosnia-Herzegovina.

41. See, for instance, Antaine Mac Coscair, "The Rajčići Operations" (unpublished manuscript).

42. "Appeals Chamber Acquits and Orders Release of Ante Gotovina and Mladen Markač" (press release), United Nations International Criminal Tribunal for the Former Yugoslavia, 16 November 2012, http://www.icty.org/en/press/appeals -chamber-acquits-and-orders-release-ante-gotovina-and-mladen-marka %C4%8D.

43. Krott, *Save the Last Bullet*, 146.

44. Demmers, "New Wars and Diasporas," 17.

45. Peter Strandberg, "Warrior Woman," *The Age*, 30 January 2004, http://www .theage.com.au/articles/2004/01/29/1075340765517.html?from=storyrhs.

46. To mention but one example, the Durand Line, which was drawn up in the late nineteenth century and now separates Afghanistan and Pakistan, cuts across ethnic Pashtun tribal areas. This had led to calls by the Afghan government and some Pashtun leaders for the creation of "Pashtunistan," a Pashtun state; if realized, this call would have significantly altered the region's borders. Elisabeth Leake, *The Defiant Border: The Afghan-Pakistan Borderlands in the Era of Decolonization, 1936–1965* (New York, 2017), esp. 104–148.

47. Cited in Riall, *Garibaldi*, 71.

48. Riall, *Garibaldi*, 71–100.

49. Lucy Riall, "Travel, Migration, Exile: Garibaldi's Global Fame," *Modern Italy* 19, no. 1 (2014): 41.

50. Gale Stokes, *Politics and Development: The Emergence of Political Parties in Nineteenth-Century Serbia* (Durham, NC, 1990), 127; and Christopher Clark, *The Sleepwalkers: How Europe Went to War in 1914* (London, 2013), 24.

51. Cited in Stokes, *Politics and Development*, 89.

52. Stokes, *Politics and Development*, 90; Clark, *The Sleepwalkers*, 26.

53. Daniel Neep, *Occupying Syria under the French Mandate: Insurgency, Space and State Formation* (New York, 2012), 51.

54. Philip S. Khoury, "Factionalism among Syrian Nationalists during the French Mandate," *International Journal of Middle East Studies* 13, no. 4 (1981): 464, 469.

55. Laila Parsons, "Rebels without Borders: Southern Syria and Palestine, 1919–1936," in *The Routledge Handbook of the History of the Middle East Mandates*, ed. Cyrus Schayegh and Andrew Arsan (London, 2015), 385–408.

56. Fauzi Al-Qawuqji, "Memoirs, 1948 Part I," *Journal of Palestine Studies* 1, no. 4 (1972): 27–58; and Laila Parsons, "Soldiering for Arab Nationalism: Fawzi al-Qawuqji in Palestine," *Journal of Palestine Studies* 36, no. 4 (2007): 33–48.

57. Cited in Parsons, "Soldiering for Arab Nationalism," 39.

58. Hafez, *Suicide Bombers*, 72; Felter and Fishman, *Al-Q'aida's Foreign Fighters*, 7.

59. While the Sykes–Picot Agreement envisaged the division of the Middle East between British and French spheres of influence, the present-day map of the region is the result of later agreements such as the Treaty of Sèvres (1920), the

Treaty of Lausanne (1923), and subsequent negotiations between France and Britain, and Turkey and Iraq.

60. Cited in Hafez, *Suicide Bombers*, 72.

61. Cited in Meghan Tinsley, "ISIS's Aversion to Sykes–Picot Tells Us Much about the Group's Future Plans," *Muftah*, 23 April 2015, http://muftah.org/the-sykes -picot-agreement-isis/#.Vz7ZuZErLIW.

62. Abdel Bari Atwan, *Islamic State: The Digital Caliphate* (London, 2015), 116–117.

63. Itai Enghel, "When the Israelis Guide the Turkish Drones—They Always Hit," *Mako*, 11 January 2011, http://www.mako.co.il/tv-ilana_dayan/2010/Article -8a5192b07893b21006.htm [Hebrew]; and Itai Enghel, "On the Frontline against Daesh," *Mako*, 22 December 2014, http://www.mako.co.il/tv-ilana_dayan/2015 /Article-55b8ff8fe637a41006.htm [Hebrew]. See also Rod Thornton, "Problems with the Kurds as Proxies against Islamic State: Insights from the Siege of Ko-bane," *Small Wars & Insurgencies* 26, no. 6 (2015): 871–872; and Tzur Shezaf, *Daash: A Journey to Satan's Doorstep* (Tel Aviv, 2016), 63 [Hebrew].

64. I am grateful to Hristo V. for sharing his fieldwork notes from Iraqi Kurdistan on this matter. For more on the divergent interests of the different Kurdish enti-ties, see Thornton, "Problems with the Kurds as Proxies," 865–885; and Shezaf, *Daash*, 59–60.

65. Roman Solchanyk, "Russia, Ukraine, and the Imperial Legacy," *Post-Soviet Af-fairs* 9, no. 4 (1993): 337–365.

66. Ivan D. Loshkariov and Andrey A. Sushentsov, "Radicalization of Russians in Ukraine: From 'Accidental' Diaspora to Rebel Movement," *Southern European and Black Sea Studies* 16, no. 1 (2016): 71–90; and Steve Rosenberg, "Ukraine: Common History Pulls in Aid from West Russia," BBC News, 23 June 2014, http://www.bbc.co.uk/news/world-europe-27961934.

67. Shaun Walker, "Putin Admits Russian Military Presence in Ukraine for First Time," *Guardian*, 17 December 2015, http://www.theguardian.com/world/2015 /dec/17/vladimir-putin-admits-russian-military-presence-ukraine.

68. Tatyana Voltskaya and Daisy Sindelar, "Volunteer Now! Russia Makes It Easy to Fight in Ukraine," *Radio Free Europe / Radio Liberty*, 4 February 2015, http://www .rferl.org/content/russia-ukraine-volunteers-kremlin-easy-to-fight-/26828559 .html.

69. Georgy Pereborshchikov for *Meduza*, "Life after Ukraine: The 'Invisible' Rus-sian Fighters Struggling to Return to Normal," *Guardian*, 29 February 2016, http://www.theguardian.com/world/2016/feb/29/ukraine-russia-fighters -donbass.

70. Leiserowitz, "Polish Volunteers in the Napoleonic Wars," 61.

71. Cited in Simon Sarlin, "Fighting the Risorgimento: Foreign Volunteers in Southern Italy (1860–1863)," *Journal of Modern Italian Studies* 14, no. 4 (2009): 482.

72. Cited in Enrico Acciai, *Antifascismo, volontariato e guerra civile in Spagna* (Milan, 2016), 57.

73. Cited in Acciai, *Antifascismo, volontariato e guerra civile*, 213.

74. Josie McLellan, "'I Wanted to Be a Little Lenin': Ideology and the German International Brigade Volunteers," *Journal of Contemporary History* 41, no. 2 (2006): 291.

75. Cited in McLellan, "'I Wanted to Be a Little Lenin,'" 291.

76. Michael Seidman, *La Victoria Nacional* (Madrid, 2012), 289; Keene, *Fighting for Franco*, 7; Hugh Thomas, *The Spanish Civil War* (London, 2003), 937–938.

77. "Notes on the Situation in the International Units in Spain. Report by Colonel Com. Sverchevsky (Walter)," 14 January 1938, in *Spain Betrayed: The Soviet Union in the Spanish Civil War*, ed. Ronald Radosh, Mary R. Habeck, and Grigory Sevostinov (New Haven, CT, 2001), 438.

78. Keene, *Fighting for Franco*, 189–194.

79. Cited in Keene, *Fighting for Franco*, 206.

80. Guevara, *The African Dream*, 51.

81. Guevara, *The African Dream*, 194.

82. Shaun Walker, "Murder in Istanbul: Kremlin's Hand Suspected in Shooting of Chechen," *Guardian,* 10 January 2016, http://www.theguardian.com/world/2016/jan/10/murder-istanbul-chechen-kremlin-russia-abdulvakhid-edelgireyev.

83. Cited in Lizzie Dearden, "Chechen Isis Fighters under Omar al-Shishani Threaten to Take Fight to Putin," *Independent,* 10 October 2014, http://www.independent.co.uk/news/world/middle-east/chechen-isis-fighters-under-omar-al-shishani-threaten-to-take-fight-to-putin-9787809.html.

84. Walker, "Murder in Istanbul."

85. Andrew E. Kramer, "Islamic Battalions, Stocked with Chechens, Aid Ukraine in War with Rebels," *New York Times,* 7 July 2015, http://www.nytimes.com/2015/07/08/world/europe/islamic-battalions-stocked-with-chechens-aid-ukraine-in-war-with-rebels.html.

86. Maria Tsvetkova, "Chechens Loyal to Russia Fight alongside East Ukraine Rebels," Reuters, 10 December 2014, http://uk.reuters.com/article/us-ukraine-crisis-chechen-fighters-idUSKBN0JO0OP20141210.

5. Controlling the Flow

1. Anthony D. Smith, *Nationalism and Modernism* (London, 2003), 20.

2. Smith, *Nationalism and Modernism*, 20.

3. For the full text of the proclamation, see George Washington, "Proclamation 4—Neutrality of the United States in the War Involving Austria, Prussia, Sardinia, Great Britain, and the United Netherlands against France," 22 April 1793,

American Presidency Project website, http://www.presidency.ucsb.edu/ws/index.php?pid=65475&st=&st1.

4. Jefferson to Genêt, 5 June 1793, in *A Message of the President of the United States to Congress Relative to France and Great Britain* (Philadelphia, 1793), 23.

5. David Riesman Jr., "Legislative Restrictions on Foreign Enlistment and Travel," *Columbia Law Review* 40, no. 5 (1940): 799.

6. Janice E. Thomson, "State Practices, International Norms, and the Decline of Mercenarism," *International Studies Quarterly* 34, no. 1 (1990): 23.

7. Riesman, "Legislative Restrictions," 801.

8. Cited in Nir Arielli, Gabriela Frei, and Inge Van Hulle, "The Foreign Enlistment, International Law, and British Politics, 1819–2014," *International History Review* 38, no. 4 (2016): 638.

9. "Officers and Crew of the Alabama," 2 April 1863, TNA, HO 45 / 7261, Part 1; and Arielli, Frei, and Van Hulle, "The Foreign Enlistment," 643–645.

10. MacKenzie, "Panslavism in Practice," 279.

11. Moss, *Russia in the Age of Alexander II*, 179–181.

12. Cited in A. N. Wilson, *Tolstoy* (London, 1988), 286.

13. Cited in Barbara Jelavich, *Russia's Balkan Entanglements 1806–1914* (Cambridge, 1991), 172.

14. Hazel Hutchison, *The War That Used Up Words: American Writers in the First World War* (New Haven, CT, 2015), 7–10.

15. James Norman Hall and Charles Bernard Nordhoff, *The Lafayette Flying Corps,* vol. 1 (Boston, 1920), 3. For more on this group of volunteers, see Philip M. Flammer, *The Vivid Air: The Lafayette Escadrille* (Athens, GA, 2008).

16. Riesman, "Legislative Restrictions," 806.

17. "Law Amending the Criminal Code of Bosnia and Herzegovina," *Official Gazette of Bosnia Herzegovina,* 47 / 14, Office of the High Representative website, http://www.ohr.int/ohr-dept/legal/laws-of-bih/pdf/New2015/BH%20Law%20Amending%20the%20CC%2047-14.pdf. The passing of this legislation did not halt the participation of Bosnians in the war in the Middle East. In March 2016 the authorities in Sarajevo believed that seventy Bosnian citizens were still engaged in the conflict in Syria. See "Na stranim ratištima trenutno boravi 70 državljana BiH, a do sada ih je poginulo 50," *Klix,* 30 March 2016, http://www.klix.ba/vijesti/bih/na-stranim-ratistima-trenutno-boravi-70-drzavljana-bih-a-do-sada-ih-je-poginulo-50/160330008.

18. The State of Israel against Ahmed Shurbaji, criminal case 22760-05-14, Haifa Magistrate Court, 3 November 2014; and "Israeli Arab Convicted in Haifa Court after Training with Islamic State in Syria," *Jerusalem Post,* 23 September 2014, http://www.jpost.com/Israel-News/Israeli-Arab-convicted-in-Haifa-court-after-training-with-Islamic-State-in-Syria-376184.

19. Miller, "IS Rape and Torture of Yezidi Women."

20. Carl-Gustaf Scott, "The Swedish Left's Memory of the International Brigades and the Creation of an Anti-Fascist Post-War Identity," *European History Quarterly* 39 (2009): 217–240; and "Treatment of Foreign Fighters in Selected Jurisdictions: Part II. Country Surveys," Library of Congress website, http://www.loc.gov/law/help/foreign-fighters/country-surveys.php#_ftnref277.

21. Cox, *The Red Army Moves,* 107.

22. Sweden did, however, agree to give Finland a number of Gloster Gladiator aircraft once all the Swedish markings had been removed. See the correspondence between Tanner and Stockholm, 1 and 2 January 1940, Finnish Foreign Ministry Archives, signum 109, folder 16.

23. Cox, *The Red Army Moves,* 105.

24. Sprague, *Swedish Volunteers,* 64, 72.

25. Martin Gutmann, *Building a Nazi Europe: The SS's Germanic Volunteers* (Cambridge, 2017), 109, 115.

26. Gutmann, *Building a Nazi Europe,* 198.

27. Winks, *The Civil War Years,* 182–190.

28. Erik Frohn-Nielsen, "Canada's Foreign Enlistment Act: Mackenzie King's Expedient Response to the Spanish Civil War" (MA thesis, University of British Columbia, 1982).

29. Cited in Petrou, *Renegades,* 54–55.

30. Cited in Petrou, *Renegades,* 55.

31. Penslar, *Jews and the Military,* 201.

32. "Government, Military Send Mixed Messages on Canadians Joining Kurds in ISIS Fight," CBS News, 22 November 2014, http://www.cbc.ca/news/world/government-military-send-mixed-messages-on-canadians-joining-kurds-in-isis-fight-1.2846120. Canada's approach toward those who fought for a foreign state or a non-state armed group in a conflict against Canada is much more robust. Bill C-24, Strengthening Canadian Citizenship Act of June 2014, provides for the possibility of stripping the Canadian citizenship of convicted dual nationals. Sandra Krähenmann, "The Obligations under International Law of the Foreign Fighter's State of Nationality or Habitual Residence, State of Transit and State of Destination," in *Foreign Fighters under International Law and Beyond,* ed. Andrea de Guttry, Francesca Capone, and Christophe Paulussen (The Hague, 2016), 248.

33. Arielli, Frei, and Van Hulle, "The Foreign Enlistment," 640, 645, 651.

34. Allan Cunningham, "The Philhellenes, Canning and Greek independence," *Middle Eastern Studies* 14, no. 2 (1978): 170.

35. Robson, " 'Strangers, Mercenaries,' " 182–184.

36. Foreign Office minutes, signed by C. E. Shuckburgh, 14 December 1936, TNA, FO 371 / 20589.

37. On the rare occasions where offenses under the Foreign Enlistment Act were brought to court, those prosecuted cannot be classed as foreign volunteers. In late December 1895 Leander Starr Jameson, the administrator of the British South Africa Company, led a privately organized force from the Cape Colony into the South African Republic with the objective of sparking a local rebellion. The Jameson Raid, as it soon became known, was a debacle. The invading force was quickly surrounded and surrendered in early January 1896. Released into British custody, Jameson and some of his men were put on trial in London later that year. They had breached section 11 of the Foreign Enlistment Act of 1870, which forbids "any person within the limits of Her Majesty's dominions, and without the license of Her Majesty," from preparing or fitting out "any naval or military expedition to proceed against the dominions of any friendly state." For more on the raid and on the subsequent proceedings see "Queen's Bench: R v Leander Starr Jameson, Sir John Christopher Willoughby Bt, Henry Frederick White, Raleigh Grey and Charles John Coventry," 1896, TNA, TS 36 / 102; and James Ruddell, "Deceit in the Transvaal? The Case of Burrows v Rhodes and Jameson," *Auckland University Law Review* 19 (2013): 99–119.

38. S. P. MacKenzie, "The Foreign Enlistment Act and the Spanish Civil War, 1936–1939," *Twentieth Century British History* 10, no. 1 (1999): 60–62; and Baxell, *British Volunteers in the Spanish Civil War*, 11.

39. Cited in Roberts, "Freedom, Faction, Fame and Blood," 153.

40. Markku Ruotsila, *Churchill and Finland: A Study in Anticommunism and Geopolitics* (London, 2005), 90–91.

41. Roberts, "Freedom, Faction, Fame and Blood," 152, 166; and Hansard, House of Commons Debate, "Finland (British Volunteers)," 14 February 1940, vol. 357 cc772-3W.

42. Margaret Thatcher, "House of Commons Statement (British Mercenaries [Angola])," 10 February 1976, Margaret Thatcher Foundation website, http://www.margaretthatcher.org/document/102956.

43. Krähenmann, "The Obligations under International Law," 242–244.

44. "Iraq Crisis: PM Urges UK Kurds Not to Travel to Fight IS," BBC News, 3 September 2014, http://www.bbc.co.uk/news/uk-29038981.

45. Robert Stradling, *The Irish and the Spanish Civil War 1936–39: Crusades in Conflict* (Manchester, 1999), 90–91.

46. "Spanish Civil War (Non-Intervention) Act, 1937," Irish Statute Book website, http://www.irishstatutebook.ie/eli/1937/act/1/enacted/en/html.

47. For estimates of the total number of Irish volunteers in the British armed forces, see Steven O'Connor, "The Pleasure Culture of War in Independent Ireland, 1922–1945," *War in History* 22, no. 1 (2015): 69.
48. Steven O'Connor, *Irish Officers in the British Forces, 1922–45* (Basingstoke, 2014), 111.
49. For more on the use of the term "Texian," see Stephen L. Hardin, *Texian Iliad: A Military History of the Texas Revolution, 1835–1836* (Austin, TX, 1994).
50. George Elliot, "Georgia and the Texas Revolution," *Georgia Historical Quarterly* 28 (1944): 1–18; Josefina Zoraida Vazquez and Lorenzo Meyer, *The United States and Mexico* (Chicago, 1987), 35; and Malet, *Foreign Fighters,* 58–88.
51. The Nationality Act, 76th Congress, 3rd session, Ch. 876, 14 October 1940.
52. Riesman, "Legislative Restrictions," 808; and "Americans Free to Join Fighting," *Sweetwater Reporter,* 26 January 1940, 1.
53. David Alan Johnson, *Yanks in the RAF* (Amherst, NY, 2015), 18–19, 258; "The Eagle Takes the Air," *Flight,* 17 October 1940, 331; Home Office to Churchill, "Formation of an Irish Brigade," 22 October 1941, TNA, PREM 3 / 129 / 5; and O'Connor, *Irish Officers in the British Forces,* 111.
54. Henderson to Marshall, "Suspension of Exports of Arms and Ammunition to Arab States and Palestine," 10 November 1947, in *Foreign Relations of the United States, 1947: The Near East and Africa,* vol. 5, 1249, http://digicoll.library.wisc.edu /cgi-bin/FRUS/FRUS-idx?id=FRUS.FRUS1947v05.
55. Livingston, *No Trophy, No Sword,* 39–47, 136; and Eric Lichtblau, "Jailed for Aiding Israel, but Pardoned by Bush," *New York Times,* 23 December 2008, http://www .nytimes.com/2008/12/24/washington/24pardons.html?_r=0.
56. "The American Citizenship of Volunteers in the Israeli Army Will Not Be Denied," *Hazofe,* 25 March 1949 [Hebrew].
57. John Casparis, "The Swiss Mercenary System: Labor Emigration from the Semiperiphery," *Review* 5, no. 4 (1982): 593–642.
58. "Code pénal militaire," 13 June 1927, Le Conseil fédéral website, https://www .admin.ch/opc/fr/classified-compilation/19270018/index.html.
59. Daniele Mariani, "No Pardon for Spanish Civil War Helpers," SWI Swissinfo .ch, 27 February 2008, http://www.swissinfo.ch/eng/no-pardon-for-spanish -civil-war-helpers/6445388.
60. Consul in Bern to Finnish Foreign Ministry, 8 December 1940, 149 / 1630 KD42, Finnish Foreign Ministry Archives, signum 109, folder 14.
61. Gutmann, *Building a Nazi Europe,* 199. For more on Riedweg, see Renat Kuenzi, "A Swiss in the Service of the SS," 4 February 2011, http://www.swissinfo.ch /eng/a-swiss-in-the-service-of-the-ss/29419904. Switzerland's strict approach of not tolerating foreign volunteering persists to the present day. In March 2015 a 33-year-old man was arrested after returning from the Middle East, where he

had joined the Syriac Military Council to fight against the Islamic State. See Henry Tuck, Tanya Silverman, and Candace Smalley, "'Shooting in the Right Direction': Anti-ISIS Foreign Fighters in Syria and Iraq," Institute for Strategic Dialogue, Horizon Series 1 (August 2016), 46–47.

62. For more on Stalin's mistreatment of non-Russian nationalities, see Isabelle Kreindler, "The Soviet Deported Nationalities: A Summary and an Update," *Soviet Studies* 38, no. 3 (1986): 387–405.

63. Benjamin Pinkus, "Change and Continuity in Soviet Policy towards Soviet Jewry and Israel, May–December 1948," *Israel Studies* 10, no. 1 (2005): 96–123. See also Golda Meir, *My Life* (London, 1975), 201–209.

64. For the full text of The Hague Convention V, signed 18 October 1907, see the Peace Resource Center website, https://www1.umn.edu/humanrts/peace /docs/con5.html.

65. Sprague, *Swedish Volunteers,* 66, 72. See also Brownlie, "Volunteers and the Law of War and Neutrality." Brownlie complained about the "inability to find conclusive evidence of governmental indifference or even activity in such recruitment."

66. Buchanan, *The Impact of the Spanish Civil War,* 3, 21, 202n101.

67. Rémi Skoutelsky, "L'engagement des volontaires français en Espagne républic-aine," *Le Mouvement social* 181 (1997): 10.

68. Riesman, "Legislative Restrictions," 819–820.

69. Thomas, *The Spanish Civil War,* 563–564; and Alpert, *A New International History of the Spanish Civil War,* 113–116.

70. Interview with Robert Doyle, recorded in 1976, Imperial War Museum, oral history recording no. 806.

71. "Security Council Unanimously Adopts Resolution Condemning Violent Extremism"; emphasis in the original.

72. "Security Council Unanimously Adopts Resolution Condemning Violent Extremism."

73. Deniz Arslan, "Obama Openly Criticizes Turkey for Foreign Fighter Crossings," *Today's Zaman,* 9 June 2015; Patrick Cockburn, "War with Isis: Obama Demands Turkey Closes Stretch of Border with Syria," *Independent,* 29 November 2015, http://www.independent.co.uk/news/world/europe/war-with-isis -president-obama-demands-that-turkey-close-stretch-of-frontier-with-syria -a6753836.html.

74. Orhan Coskun and Humeyra Pamuk, "Turkey Says It Notified France Twice about Attacker, Says Senior Official," *Independent,* 16 November 2015, http://www .independent.co.uk/news/world/europe/paris-attacks-turkey-says-it-notified -france-twice-about-attacker-says-senior-official-a6736131.html.

75. "UN Says '25,000 Foreign Fighters' Joined Islamist Militants," BBC News, 2 April 2015, http://www.bbc.co.uk/news/world-middle-east-32156541.

76. Martin Chulov, Jamie Grierson and Jon Swaine, "Isis Faces Exodus of Foreign Fighters as Its 'Caliphate' Crumbles," *Guardian*, 26 April 2017, https://www.theguardian.com/world/2017/apr/26/isis-exodus-foreign-fighters-caliphate-crumbles.

77. Timothy Holman, "Why States Fail to Counter Foreign Fighter Mobilizations: The Role of Intelligence Services," *Perspectives on Terrorism* 10, no. 6 (2016), http://www.terrorismanalysts.com/pt/index.php/pot/article/view/565/html.

78. In February 2017 the UNHCR estimated the number of Syrian refugees in Turkey at more than 2.9 million. "Syria Regional Refugee Response," UNHCR website, http://data.unhcr.org/syrianrefugees/regional.php.

79. Axel Jansen, "Heroes or Citizens? The 1916 Debate on Harvard Volunteers in the 'European War,'" in *War Volunteering in Modern Times: From the French Revolution to the Second World War*, ed. Christine G. Krüger and Sonja Levsen (Basingstoke, 2011), 151.

80. Riesman, "Legislative Restrictions," 817.

81. See, for instance, Gurney, *Crusade in Spain,* 39; Haldane, *Truth Will Out,* 117.

82. "France Seizes Passports of Six 'Syria-Bound' Citizens," BBC News, 23 February 2015, http://www.bbc.co.uk/news/world-europe-31587651.

83. Gutmann, *Buidling a Nazi Europe,* 63–64.

84. Cited in MacKenzie, "The Foreign Enlistment Act," 62.

85. Johnson, *Yanks in the RAF,* 19.

86. Author's interview with Menachem Silberstein, Tel Aviv, 5 January 2011. Fellow volunteer Zeev Feliswasser, interviewed on 20 December 2010, recalled facing a similar questioning at the customs in Dover. For lists of "aliens" and British citizens who were suspected by British authorities of having gone to Palestine to enlist in the Jewish armed forces, see TNA, HO 45 / 25587.

87. Brown, *Adventuring through Spanish Colonies,* 17, 26.

88. St. Clair, *That Greece Might Still Be Free,* 120–123.

89. Vale, *Independence or Death!,* 30.

90. Sutcliffe, "British Red Shirts," 207.

91. Kelly, "This Ain't Ethiopia," 16–18.

92. Kemp, *The Thorns of Memory,* 8.

93. Haldane, *Truth Will Out,* 100–123.

94. Enrico Acciai, "I primi volontari italiani nella guerra civile spagnola: Genesi e nascita della Sezione Italiana della Colonna Ascaso," *Ebre* 38 (2010): 20; and David Wingeate Pike, *France Divided: The French and the Civil War in Spain* (Brighton, UK, 2011), 329n16.

95. Gaunt, *War and Beer,* 5–7.

96. Hutt, *Paint,* 90–92.

97. For more on the use of online social networks for recruitment purposes, see, for instance, Joseph A. Carter, Shiraz Maher, and Paul R. Neumann, "Greenbirds: Measuring Importance and Influence in Syrian Foreign Fighter Networks," ICSR Paper, 2014, http://icsr.info/wp-content/uploads/2014/04/ICSR-Report -Greenbirds-Measuring-Importance-and-Infleunce-in-Syrian-Foreign-Fighter -Networks.pdf; and Sharon Marris, "A Terrorist's Travel Guide," *Metro*, 19 June 2014, 7.

98. Daniel Byman and Jeremy Shapiro, "Be Afraid. Be a Little Afraid: The Threat of Terrorism from Western Foreign Fighters in Syria and Iraq," Foreign Policy at Brookings Policy Paper No. 34 (November 2014), https://www.brookings.edu /wp-content/uploads/2016/06/Be-Afraid-web.pdf, esp. 10–12.

99. Gurney, *Crusade in Spain*, 38.

100. Rob Stradling, "The Spies Who Loved Them: The Blairs in Barcelona, 1937," *Intelligence and National Security* 25, no. 5 (2010): 641–642.

101. Thomas, *The Spanish Civil War*, 441; and Wintringham, *English Captain*, 114.

102. Benvenuto to Jewish Agency in Rome, 1 June 1948, Central Zionist Archives in Jerusalem (CZA), L16 / 68.

103. Cuffao to Hebrew legation in Rome, 27 August 1948, CZA, L16 / 68. The file contains several other letters from individuals prepared to offer their services.

104. Josef Willner to Sandor Dusnoki, 30 November 1948; and Arie Stern to Emil Winter, 14 November 1948, CZA, L16 / 68.

105. Nir Arielli, "Recognition, Immigration and Divergent Expectations: The Reception of Foreign Volunteers in Israel during and after the Wars of 1948 and 1967," *Journal of Modern European History* 14, no. 3 (2016): 374–390.

106. Haldane, *Truth Will Out*, 117.

107. Ralph Lowenstein, "North American Volunteers in the Israeli Army," *Jewish Virtual Library*, http://www.jewishvirtuallibrary.org/jsource/History/IsraelArmy .html.

108. "Report of the United Nations Mediator on Palestine to the Security Council," 12 July 1948, United Nations Security Council website, https://unispal.un.org /DPA/DPR/unispal.nsf/0/06EB90C615E1DBCD802564B400556E40.

109. Nir Arielli, "When Are Foreign Volunteers Useful? Israel's Transnational Soldiers in the War of 1948 Re-examined," *Journal of Military History* 78, no. 2 (April 2014): 710.

110. Author's interview with Avi Grant, Ra'anana, Israel, 21 December 2010.

111. The State of Israel against Ahmed Shurbaji.

112. See, for instance, Ben Tufft, "Isis 'Executes Up to 200 Fighters' for Trying to Flee Jihad and Return Home," *Independent*, 29 December 2014, http://www .independent.co.uk/news/world/middle-east/isis-executes-at-least-120-fighters -for-trying-to-flee-and-go-home-9947805.html.

113. Centner, *From Madrid to Berlin,* 81, 137; Nahum Pondak, "We Fought for Every Rock," *Davar Hashavua,* 1 July 1966, 7; Ayelet Negev, "A War of Choice," *Yedioth Ahronoth,* 8 September 2006, 108–112; and Dan Yahav, *They Too Were Heroes: Volunteer Fighters from Eretz Israel in the International Brigades in Spain (1936–1938)* ([Tel Aviv], 2007), 70–77, 103–105, 183–189 [all sources are in Hebrew].

114. Milton Bearden, "Afghanistan, Graveyard of Empires," *Foreign Affairs,* November–December 2001, 24.

115. Shimon Prokupecz, "Denver Teens Set Out for Syria to Join Extremists; Parents, FBI Bring Them Back," CNN, 23 October 2014, http://edition.cnn.com/2014/10/21/us/colorado-teens-syria-odyssey/.

116. Brett Holman, *The Next War in the Air: Britain's Fear of the Bomber, 1908–1941* (London, 2014), 57.

6. Winning Wars?

1. "Security Council Unanimously Adopts Resolution."

2. There have been attempts to examine the impact of foreign insurgents on the duration and intractability of localized conflicts, such as the one in Chechnya between 1994 and 2005. See, for instance, Kristin M. Bakke, "Help Wanted? The Mixed Record of Foreign Fighters in Domestic Insurgencies," *International Security* 38, no. 4 (2014): 150–187.

3. Allan R. Millett, Williamson Murray, and Kenneth H. Watman, "The Effectiveness of Military Organizations," in *Military Effectiveness,* vol. 1, ed. Allan R. Millett and Williamson Murray (Boston, 1988), 2.

4. Millett, Murray, and Watman, "The Effectiveness of Military Organizations," 6.

5. Malet, *Foreign Fighters,* 4.

6. Malet, *Foreign Fighters,* 11.

7. Malet, *Foreign Fighters,* 52–53.

8. Coulombe, *The Pope's Legion;* and Ferdinand Nicolas Göhde, "A New Military History of the Italian Risorgimento and Anti-Risorgimento: The Case of 'Transnational Soldiers,'" *Modern Italy* 19, no. 1 (2014): 21–39.

9. See Jackson, *Fallen Sparrows,* 68; Rémi Skoutelsky, *L'espoir guidait leurs pas: Les volontaires français dans les Brigades internationales 1936–1939* (Paris, 1998), 330; and Thomas, *The Spanish Civil War,* 941–942.

10. See Jackson, *Fallen Sparrows,* 105; S. P. MacKenzie, *Revolutionary Armies in the Modern Era: A Revisionist Approach* (London, 1997), 132. For estimates regarding the total size of the Republican military force, see Michael Seidman, *Republic of Egos: A Social History of the Spanish Civil War* (Madison, WI, 2002), 40; and Michael Alpert, *The Republican Army in the Spanish Civil War 1936–1939* (Cambridge, 2013), ix, 82, 269.

11. Jari Leskinen and Antti Juutilainen, *Talvisodan Pikkujättiläinen* (Porvoo, 2009), 770–776; Sotatieteen Laitos, *Talvisodan Historia*, vol. 1 (Porvoo, 1977), 165; and Sotatieteen Laitos, *Talvisodan Historia*, vol. 4 (Porvoo, 1979), 52.

12. Arielli, "When Are Foreign Volunteers Useful?," 718. The exact number of Machal recruits is contested. Shortly after the war some Israeli officials believed that there had been just over 2,400 foreigners in the IDF. Decades later, historian David Bercuson put the number of foreign volunteers at more than 5,000. For more on this debate, see David Bercuson, *The Secret Army* (Toronto, 1983), xiii; and Yaacov Markovitzky, *Machal: Overseas Volunteers in Israel's War of Independence* (Tel Aviv, 2003), 7–8.

13. Maliach and Shay, *From Kabul to Jerusalem*, 44.

14. Ed Giradet, *Afghanistan: The Soviet War* (Abingdon, UK, 2011 [1985]), 34; and Lester W. Grau, "Breaking Contact without Leaving Chaos: The Soviet Withdrawal from Afghanistan," *Journal of Slavic Military Studies* 20, no. 2 (2007): 244.

15. Anthony H. Cordesman, "Iraq's Evolving Insurgency and the Risk of Civil War," Center for Strategic and International Studies, 26 April 2006, 153, https://csis-prod .s3.amazonaws.com/s3fs-public/legacy_files/files/media/csis/pubs/060424 _iraqinsurgrpt.pdf; Fawaz A. Gerges, *The Far Enemy: Why Jihad Went Global* (Cambridge, 2005), 260; Bruce Hoffman, "Insurgency and Counterinsurgency in Iraq," *Studies in Conflict & Terrorism* 29, no. 2 (2006): 111–112; and Hafez, *Suicide Bombers in Iraq*, 71.

16. Millett, Murray, and Watman, "The Effectiveness of Military Organizations," 19–26.

17. See, for instance, MacGregor Knox, *Common Destiny: Dictatorship, Foreign Policy, and War in Fascist Italy and Nazi Germany* (Cambridge, 2000), 181.

18. Robert Molis, *Les Francs-Tireurs et les Garibaldi: Soldats de la République 1870–1871 en Bourgogne* (Paris, 1995), 15, 67, 260; Karine Varley, "Contesting Concepts of the Nation in Arms: French Memories of the War of 1870–1 in Dijon," *European History Quarterly* 36, no. 4 (2006): 548–573; and Pécout, "Philhellenism in Italy," 413–414.

19. For a positive appraisal of Garibaldi's contribution, see, for instance, W. Alexander Vacca and Mark Davidson, "The Regularity of Irregular Warfare," *Parameters* 41, no. 1 (2011): 21.

20. Varley, "Contesting Concepts," 561–562; Michael Howard, *The Franco-Prussian War* (London, 1991 [1961]), 427; Ernest Alfred Vizetelly, *My Days of Adventure: The Fall of France, 1870–71* (London, 1914), 321; Riall, *Garibaldi*, 353–354; and Molis, *Les Francs-Tireurs*, 67. Much like the conservative French critics, the Italian minister in Berlin, Edoardo De Launay, was dismissive of Garibaldi's military successes. De Launay hoped his government would distance itself from the old general and downplay his achievements in order not to jeopardize relations with

Germany. De Launay to Visconti Venosta, 15 February 1871, *Documenti Diplomatici Italiani (DDI)*, series 2, vol. 2 (Rome, 1966), 179–181.

21. Vizetelly, *My Days of Adventure*, 220–226; Molis, *Les Francs-Tireurs*, 79; and Howard, *The Franco-Prussian War*, 409–410.

22. Eva Cecchinato, *Camicie rosse: I garinaldini dall'Unità alla Grande Guerra* (Rome, 2007), 142; Galateri di Genola to Visconti Venosta, 8 March 1871, *DDI*, series 2, vol. 2, 255–256; and Molis, *Les Francs-Tireurs*, 337–341.

23. Former Zouaves, who fought for France after their removal from Italy following the conquest of the Papal States, suffered higher casualty rates. Posted to Patay in December 1870, one of their battalions lost 218 men out of 300. Coulombe, *The Pope's Legion*, 188–190.

24. Hubert Heyriès, "The Garibaldian Volunteers in France during the First World War," *Journal of Modern European History* 14, no. 3 (2016): 359–373.

25. Coletti, *Peppino Garibaldi*, 111–112. Teenage volunteers who ran away to join a foreign conflict were not unique to the First World War. There were a few recorded instances of Canadian volunteers in the Union army during the American Civil War who were as young as 14 and 15. See Winks, *The Civil War Years*, 185.

26. Heyriès, "The Garibaldian Volunteers," 365–366.

27. Seidman, *Republic of Egos*, 40; "Mr. Eden's Statement," Hansard, HC Deb, 19 January 1937, vol. 319 cc92-161, http://hansard.millbanksystems.com/commons/1937/jan/19/mr-edens-statement#S5CV0319P0_19370119_HOC_428; Wintringham, *English Captain*, 115; Alpert, *The Republican Army*, 221. Canadian volunteers were, on average, older than their comrades from other countries. In 1937 their average age was 32, and the most common ages of recruits were 33–34. Petrou, *Renegades*, 13.

28. Richardson, *Comintern Army*, 85; Jackson, *Fallen Sparrows*, 104–105; MacKenzie, *Revolutionary Armies*, 118–119.

29. Wintringham, *English Captain*, 139. See also Richardson, *Comintern Army*, 81–84; Seidman, *Republic of Egos*, 59; and Helen Graham, *The Spanish Republic at War 1936–1939* (Cambridge, 2002), 177.

30. Petrou, *Renegades*, 21; Gurney, *Crusade in Spain*, 76; MacKenzie, *Revolutionary Armies*, 123.

31. R. Dan Richardson, "The Defense of Madrid: Mysterious Generals, Red Front Fighters, and the International Brigades," *Military Affairs* 43, no. 4 (1979): 178–185; and Ronald Radosh, Mary R. Habeck, and Grigory Sevostinov, eds., *Spain Betrayed: The Soviet Union in the Spanish Civil War* (New Haven, CT, 2001), 104.

32. Richardson, *Comintern Army*, 83, 87–88; Jackson, *Fallen Sparrows*, 105; MacKenzie, *Revolutionary Armies*, 132.

33. Skoutelsky, *L'espoir*, 82; Alpert, *The Republican Army*, 222; and "Notes on the Situation in the International Units in Spain: Report by Colonel Com. Sverchevsky (Walter)," 14 January 1938, in Radosh, Habeck, and Sevostinov, *Spain Betrayed*, 439. The assessment of the effectiveness of the foreign volunteers who fought on Franco's side seems more straightforward. General O'Duffy's Irish Brigade in particular performed poorly. The 700-strong unit arrived in Spain in November 1936. Its first and only engagement was on the Jarama front in February 1937. On their way there, they were fired upon by another Nationalist unit that mistook them for a force belonging to the Republican International Brigades. In the cross-fire that ensued, some five Irishmen and a larger number of Nationalist Spaniards were killed. A few days later, an engagement between the Irish Brigade and Republican forces ended in the former's rout under fire. Subsequently, the unit, whose men voted to leave Spain and return to Ireland, was disbanded. Judith Keene, "Fighting for God, for Franco and (Most of All) for Themselves: Right-Wing Volunteers in the Spanish Civil War," in *War Volunteering in Modern Times: From the French Revolution to the Second World War*, ed. Christine G. Kruger and Sonja Levsen (Basingstoke, 2011), 213, 220; and Christopher Othen, *Franco's International Brigades: Adventurers, Fascists, and Christian Crusaders in the Spanish Civil War* (London, 2013), 133–135.

34. See, for instance, Steiner, *Die Freiwilligen*.

35. See, for instance, Robert Forbes, *For Europe: The French Volunteers of the Waffen SS* (Mechanicsburg, PA, 2006).

36. Philippe Carrard, *The French Who Fought for Hitler: Memories from the Outcasts* (Cambridge, 2010), 21–24; MacKenzie, *Revolutionary Armies*, 134–141; and Gutmann, *Building a Nazi Europe*, 1–2, 173.

37. Owen Anthony Davey, "The Origins of the Legion des Volontaires Français contre le Bolchevisme," *Journal of Contemporary History* 6, no. 4 (1971): 29–45.

38. Carrard, *The French Who Fought for Hitler*, 12–13; and Kenneth W. Estes, *A European Anabasis: Western European Volunteers in the German Army and SS, 1940–1945*, e-book ed. (New York, 2008), paras. 357–358.

39. Cited in Forbes, *For Europe*, 457.

40. Estes, *A European Anabasis*, para. 358; and MacKenzie, *Revolutionary Armies*, 145–146.

41. Gutmann, *Building a Nazi Europe*, 146; Estes, *A European Anabasis*, para. 59.

42. MacKenzie, *Revolutionary Armies*, 153–154.

43. Noel Malcolm, *Bosnia: A Short History* (London, 1996), 189–190; and Motadel, *Islam and Nazi Germany's War*, 229, 310.

44. Ernest Hemingway, *For Whom the Bell Tolls* (London, 2004 [1940]), 141–142.

45. Graciela Iglesias Rogers, "Bewilderment, Gratitude and Contempt: Hosting Foreign Volunteers during the Napoleonic Wars," paper delivered at the workshop Transnational Encounters, University of Leeds, June 2014.

46. Heyriès, "The Garibaldian Volunteers," 364.

47. Gurney, *Crusade in Spain*, 96–97.

48. Author's interviews with Sol Jacobs, Tel Ganim, Israel, 22 December 2010, and Lily Myerson, Nordea, Israel, 23 December 2010.

49. Penslar, *Jews and the Military*, 233, 236.

50. See, for instance, Bergen, *The Osama bin Laden I Know*, 28; and Hutt, *Paint*, 94, 105.

51. Göhde, "A New Military History," 28–29.

52. Pécout, "Philhellenism in Italy," 418–419; and "Latest Intelligence: The Cretan Question," *Times* [London], 2 April 1897, 5. Bertet, whose efforts displeased Ricciotti Garibaldi, struggled to receive the support of the Greek government.

53. Leskinen and Juutilainen, *Talvisodan Pikkujättiläinen*, 773–775.

54. Barak Mendelsohn, "Foreign Fighters—Recent Trends," *Orbis* 55, no. 2 (2011): 190; and Timothy B. McCulloh, "The Inadequacy of Definition and the Utility of a Theory of Hybrid Conflict: Is the Hybrid Threat New?," in *Hybrid Warfare and the Gray Zone Threat*, ed. Douglas C. Lovelace (Oxford, 2016), 64.

55. Brown, *Adventuring through Spanish Colonies*, 62, 68, 82.

56. Brown, *Adventuring through Spanish Colonies*, 40–42, 61–64.

57. Leventhal, *The Last Dissenter*, 30–32.

58. Cited in Macnab, *The French Colonel*, 15.

59. Pretorius, "Welcome but Not That Welcome," 127–133.

60. Haim Levenberg, *Military Preparations of the Arab Community in Palestine 1945–1949* (London, 1993), 199–209; and Benny Morris, *1948: The First Arab-Israeli War* (New Haven, CT, 2008), 90–93, 133–138.

61. Qawuqji, "Memoirs, 1948 Part I," 48.

62. See, for instance, "Reports from 'Gardner,'" Nablus, 14 March 1948, Haganah Archives, 105 / 216: "Military Activity from the Arab Countries."

63. "Dafna" to "Tzuriel," 15 April 1948, IDF Archives, 8275 / 1949, file 136; Hillel Cohen, *Army of Shadows: Palestinian Collaboration with Zionism* (Berkley, 2008), 233–234, 243, 251; and Itamar Radai, *A Tale of Two Cities: The Palestinian Arabs in Jerusalem and Jaffa, 1947–1948* (Tel Aviv, 2015), 59, 128–132 [Hebrew].

64. Qawuqji, "Memoirs, 1948 Part I," 58.

65. Cited in Bergen, *The Osama bin Laden I Know*, 42.

66. Brian Glyn Williams, "Return of the Arabs: Al-Qa'ida's Current Military Role in the Afghan Insurgency," *CTC Sentinel* 1, no. 3 (2008): 22–25.

67. See Maliach and Shay, *From Kabul to Jerusalem*, 51–53; and Peter Bergen and Paul Cruickshank, "Revisiting the Early Al Qaeda: An Updated Account of its Formative Years," *Studies in Conflict and Terrorism* 35 (2011): 5–6.

68. Mustafa Hamid and Leah Farrall, *The Arabs at War in Afghanistan* (London, 2015), 98.

69. Cited in Bergen, *The Osama bin Laden I Know,* 88. See also Maliach and Shay, *From Kabul to Jerusalem,* 57–60; and Bergen and Cruickshank, "Revisiting the Early Al Qaeda," 9.

70. Krott, *Save the Last Bullet,* 11, 68; author's correspondence with Ivan Farina, August 2010; and Hutt, *Paint,* 95, 108, 189.

71. Author's correspondence with Antaine Mac Coscair, June 2010 and March 2016; Antaine Mac Coscair, "John 'Olly' Scarrott," http://cascarino.homestead.com/olly.html; and Mac Coscair's filmed interview with Miša Ciprijanović, Rajić, 19 October 2007.

72. Cited in Zipporah Porath, *Col. David (Mickey) Marcus* (New York, 2010), 14.

73. Krott, *Save the Last Bullet,* 19–33; Arielli, "In Search of Meaning," 13.

74. Bakke, "Help Wanted?," 167.

75. Mendelsohn, "Foreign Fighters," 198; Kohlmann, *Al-Qaida's Jihad in Europe,* 16–24.

76. Fausto Biloslavo, "Ukraine: Far-Right Fighters from Europe Fight for Ukraine," Eurasianet.org, 6 August 2014, http://www.eurasianet.org/node/69401.

77. Murray, *Wings over Poland,* 11.

78. Murray, *Wings over Poland,* 24–25.

79. Murray, *Wings over Poland,* 243.

80. Cited in Murray, *Wings over Poland,* 349.

81. A recent study on the conflict concludes that "the Poles won the war in large part because of errors committed by the Soviets." Jerzy Borzęcki, "Battle of Warsaw, 1920: Was Radio Intelligence the Key to Polish Victory over the Red Army?," *Journal of Military History* 81, no. 2 (2017): 468.

82. Thon to Ben Gurion, 28 January 1949, IDF Archive, 580 / 1956, file 240; Arielli, "When Are Foreign Volunteers Useful?," 705–706; Bercuson, *The Secret Army,* 34–38; and Livingston, *No Trophy, No Sword,* 78.

83. Morris, *1948,* 85, 206–207; and United Nations Security Council, Resolution 50 of 29 May 1948, S / 801, https://unispal.un.org/DPA/DPR/unispal.nsf/0/6B76F035CD9C4A36852560C200599BB7.

84. Author's interviews with Sol Jacobs and Anita Koifman, Tel Ganim, Israel, 22 December 2010.

85. Amitzur Ilan, *The Origins of the Arab-Israeli Arms Race: Arms, Embargo, Military Power and Decisions in the 1948 Palestine War* (London, 1996), 5–6.

86. Stephen Biddle, Wade Hinkle, and Michael Fischerkeller, "Skill and Technology in Modern Warfare," *Joint Force Quarterly* 22 (1999): 18–27.

87. Dave Kilcullen, "Anatomy of a Tribal Revolt," *Small Wars Journal Blog,* 29 August 2007, http://smallwarsjournal.com/blog/anatomy-of-a-tribal-revolt.

88. Hafez, *Suicide Bombers in Iraq,* 71–76.

89. Felter and Fishman, *Al-Q'aida's Foreign Fighters,* 18–19.

90. Mendelsohn, "Foreign Fighters," 197.

91. Hafez, *Suicide Bombers in Iraq*, 103–104.

92. Cited in Cordesman, "Iraq's Evolving Insurgency," 129.

93. Suicide attacks, which were introduced to the conflict in Chechnya in 2000, most likely by foreign fighters, had never been part of the tradition of Chechen resistance. As Bakke illustrates, this tactical innovation was resented by the majority of the local population as well as by key resistance leaders. Bakke, "Help Wanted?," 179–184.

94. Terzuolo, "The Garibaldini in the Balkans," 120–122.

95. Heyriès, "The Garibaldian Volunteers," 369.

96. Sandes, *The Autobiography of a Woman Soldier*, 13.

97. *Salaria Kea: A Negro Nurse*. The deeds of American volunteers who fought and died in Spain were celebrated in a number of contemporary pamphlets and publications. See, for instance, Otis Hood and Phil Frankfeld, *Americans in Spain: New England fights for Spanish Democracy* (pamphlet) [n.d.], ALBA, VF, box 5, folder 64, "New England Volunteers."

98. Regler, *The Owl of Minerva*, 315.

99. See, for instance, Seidman, *Republic of Egos*, 59; Graham, *The Spanish Republic at War*, 177.

100. The English translation of the text is taken from "La Pasionaria's Farewell Address," Modern American Poetry website, http://www.english.illinois.edu/maps/scw/farewell.htm.

101. Nicholas John Cull, *Selling War: The British Propaganda Campaign against American Neutrality in World War II* (New York, 1995), 89–90, 181–182.

102. Mendelsohn, "Foreign Fighters," 199.

103. Adam Goldman and Eric Schmitt, "One by One, ISIS Social Media Experts Are Killed as Result of F.B.I. Program," *New York Times*, 24 November 2016, https://www.nytimes.com/2016/11/24/world/middleeast/isis-recruiters-social-media.html?_r=0.

104. Cited in Adam Lucente, "The YPGs Foreign Fighters: 'Western Face on a Foreign Problem,'" *News Deeply*, 26 August 2015, https://www.newsdeeply.com/syria/articles/2015/08/26/the-ypgs-foreign-fighters-western-face-on-a-foreign-problem.

105. Graham, *The Spanish Republic at War*, 139–150; Seidman, *Republic of Egos*, 55; Alpert, *The Republican Army in the Spanish Civil War*, 65–70, 81–83, 222.

106. Victoria Clark, "The War According to Shane, Irishman for a Free Croatia," *Observer*, 24 November 1991, 19; author's correspondence with Antaine Mac Coscair, June 2010 and March 2016.

107. Marko Attila Hoare, *How Bosnia Armed* (London, 2004), 77–80, 102–107.

108. Denys Krasnikov, "Rada Lets Foreigners Serve in the Ukrainian Army," *Kyiv-Post*, 6 October 2015, http://www.kyivpost.com/article/content/ukraine/rada-lets-foreigners-serve-in-ukrainian-army-399418.html.

109. Bakke, "Help Wanted?," 177, 185–187.

110. Laia Balcells and Stathis Kalyvas, "Technology of Rebellion in the Syrian Civil War," in *The Political Science of Syria's War* (2013), 11–12, https://pomeps.org/wp-content/uploads/2013/12/POMEPS_BriefBooklet22_PoliSciSyria_Web.pdf.

111. See Richardson, *Comintern Army*, 81–89; MacKenzie, *Revolutionary Armies*, 117–118; Seidman, *Republic of Egos*, 87.

112. For an early attempt to take stock of the situation, see Fabrizio Coticchia, "The Military Impact of Foreign Fighters on the Battlefield: The Case of the ISIL," in *Foreign Fighters under International Law and Beyond*, ed. Andrea de Guttry, Francesca Capone, and Christophe Paulussen (The Hague, 2016), 121–140.

7. The Dark Side

1. Leo Tolstoy, *Anna Karenina* (London, 1999), 762.

2. Wilson, *Tolstoy*, 286–287; Moss, *Russia in the Age of Alexander II*, 179.

3. For instance, Garibaldi described the pope's foreign volunteers, the Zouaves, as "foreign cowards" and mercenaries. Coulombe, *The Pope's Legion*, 123.

4. "Inside Story: Dogs of War," *BBC1*, 20 May 1992.

5. Letter by Joseph Dallet, 19 March 1937, ALBA, Joseph Dallet collection, box 1, folder 11, correspondence, 1937.

6. Cited in Kai Bird and Martin J. Sherwin, *American Prometheus: The Triumph and Tragedy of J. Robert Oppenheimer* (London, 2009), 159.

7. St. Clair, *That Greece Might Still Be Free*, 83–85, 89–90; Gordon, *History of the Greek Revolution*, 431.

8. St. Clair, *That Greece Might Still Be Free*, 90–96.

9. Burbury to Gripenberg, 4 February 1940, Finnish Foreign Ministry Archives, signum 109, folder 14.

10. Clipping sent from Stockholm: "Spanish Airman to Aid the Finns in Fighting Former Ally" [n.d.], Finnish Foreign Ministry Archives, signum 109, folder 14.

11. Suontausta (Copenhagen) to the Finnish Foreign Ministry, 29 April 1940, Finnish Foreign Ministry Archives, signum 109, folder 14.

12. John Sweeney, "Who Killed Paul Jenks?" and "The Killer Who Loves ET and Marry Poppins," *Observer*, 26 June 1994, 12–15. Würtenberg was allegedly killed because he was suspected of being a spy, while Jenks had been asking questions about Würtenberg's death.

13. For some useful commentary on and links to both films, see Dave Cinzano's blog: *Balkan Scrapbook: Cuttings, Clippings, Photos & Films from the Yugoslav Civil War*, https://balkanscrapbook.wordpress.com/.

14. For more on his exploits in Surinam and elsewhere, see Karl Penta, *A Mercenary Tale* (London, 2002); and Karl Penta, *Have Gun Will Travel* (London, 2003).

15. Krott, *Save the Last Bullet*, 201–202. Rózsa-Flores depicted his own, slightly more positive account of his wartime service in Croatia in the autobiographical film *Chico* (2001). Rózsa-Flores was killed by police in Bolivia in 2009 for allegedly plotting to assassinate President Evo Morales.

16. Gaunt, *War and Beer*, 51.

17. Krott, *Save the Last Bullet*, 141, 155.

18. Krott, *Save the Last Bullet*, 154, 209.

19. Krott, *Save the Last Bullet*, 155, 214.

20. Krott, *Save the Last Bullet*, 154.

21. Krott, *Save the Last Bullet*, 178.

22. Krott, *Save the Last Bullet*, 212.

23. For English translations of excerpts from the book and a discussion of Casagrande's role in the war in Bosnia and his later attempts to distance himself from this episode, see "Marco Casagrande—The Mostar Road Hitchhiker," *Johnnyantora: Dispatches from War and Culture* (blog), 7 November 2013, http://johnnyantora.com/2013/11/07/marco-casagrande-the-mostar-road-hitchhiker/.

24. "Suomalaisen palkkasoturin teot taas viranomaisten syyniin," *Ilta-Sanomat*, 9 February 2001, http://www.iltasanomat.fi/kotimaa/art-2000000014761.html.

25. Author's correspondence with Roland Bartetzko, July 2015.

26. Paul Britton, "Former Soldier Who 'Flew to Syria to Fight ISIS' While on the Run Is Arrested at Manchester Airport," *Manchester Evening News*, 26 November 2015, http://www.manchestereveningnews.co.uk/news/greater-manchester-news/former-soldier-who-flew-syria-10507741.

27. Daniel H. Heinke, "ICSR Insight—German Jihadists in Syria and Iraq: An Update," International Centre for the Study of Radicalisation, 29 February 2016, http://icsr.info/2016/02/icsr-insight-german-jihadists-syria-iraq-update/.

28. Anne Penketh, "French Suspect in Brussels Jewish Museum Attack Spent Year in Syria," *Guardian*, 1 June 2014, http://www.theguardian.com/world/2014/jun/01/french-suspect-brussels-jewish-museum-attack-syria; Jacques Follorou, "Mehdi Nemmouche, geôlier d'otages occidentaux en Syrie," *Le Monde*, 6 September 2014, http://www.lemonde.fr/international/article/2014/09/06/mehdi-nemmouche-geolier-des-ex-otages-en-syrie_4483126_3210.html; and David Chazan, "Brussels Museum Shooting Suspect 'Beheaded Baby,'" *Telegraph*, 7 September 2014, http://www.telegraph.co.uk/news/worldnews/middleeast/syria/11080079/Brussels-museum-shooting-suspect-beheaded-baby.html.

29. St. Clair, *That Greece Might Still Be Free,* 83, 85, 100–102.

30. Hamid and Farrall, *The Arabs at War in Afghanistan,* 36.

31. Maliach and Shay, *From Kabul to Jerusalem,* 38, 47–49.

32. Author's correspondence with Antaine Mac Coscair, June 2010.

33. Hutt, *Paint,* 105.

34. Hutt, *Paint,* 106.

35. Hutt, *Paint,* 121.

36. Author's correspondence with Roland Bartezko, July 2015 and April 2016.

37. See, for instance, "'Lessons Learned'—By Mujahid Commander."

38. Jeremy Scahill, "The Purge: How Somalia's Al-Shabaab Turned against Its Own Foreign Fighters," *Intercept,* 19 May 2015, https://theintercept.com /2015/05/19/somalia-al-shabaab-foreign-fighter-cia/; and "8 Reasons Why al-Shabaab Killed al-Amriki," *Sahan Journal,* 28 September 2013, http://sahan journal.com/al-shabaab-killed-al-amriki-omar-hammami/#.VwfSeK R97IU.

39. Robert Young Pelton, "The All-American Life and Death of Eric Harroun," *Vice News,* 11 April 2014, https://news.vice.com/article/the-all-american-life-and -death-of-eric-harroun; and "The American Jihadist: Eric Harroun in His Own Words," *Vice News,* 13 April 2014, https://news.vice.com/video/american-jihadist -eric-harroun-in-his-own-words.

40. St. Clair, *That Greece Might Still Be Free,* 122–125.

41. Gordon, *History of the Greek Revolution,* 485.

42. Cited in Jelavich, *Russia's Balkan Entanglements,* 170.

43. Cited in MacKenzie, "Panslavism in Practice," 284.

44. Cited in MacKenzie, "Panslavism in Practice," 288.

45. Geoffrey Swain, "The Disillusioning of the Revolution's Praetorian Guard: The Latvian Riflemen, Summer–Autumn 1918," *Europe-Asia Studies* 51, no. 4 (1999): 667–686. According to intelligence reports compiled by the Whites, Latvian troops were used not only to suppress strikes but also to carry out summary executions and discipline Red Army recruits. See Whitewood, "Nationalities in a Class War," 357.

46. Volgyes, "Hungarian Prisoners of War in Russia," 83–84.

47. "Bela Kun's Arrest in Russia Reported," *New York Times,* 3 August 1937, 10.

48. Volgyes, "Hungarian Prisoners of War in Russia," 84–85.

49. See Fitzroy Maclean, *Disputed Barricade: The Life and Times of Josip Broz* (London, 1957), 23–27; Phyllis Auty, *Tito: A Biography* (Harlow, UK, 1970), 33–38; Stevan K. Pavlowitch, *Tito: Yugoslavia's Great Dictator: A Reassessment* (London, 2006), 12–14; Neil Barnett, *Tito* (London, 2006), 25–29; and Geoffrey Swain, *Tito: A Biography* (London, 2011), 8–9.

50. Jackson, *"For Us It Was Heaven,"* 100.

51. Many of the Yugoslav volunteers who arrived in Spain did not come directly from Yugoslavia but from France, Belgium, and other countries where they had been working, most commonly as miners.

52. Risto Miljković, *Španija 1936–1939: Jugosloveni bivši dobrovoljci španske republikanske vojske* (Belgrade, 1970), 23–26; Phyllis Auty, "Popular Front in the Balkans: 1. Yugoslavia," *Journal of Contemporary History* 5, no. 3 (1970): 57–58; Vjeran Pavlaković, "Twilight of the Revolutionaries: 'Naši Španci' and the End of Yugoslavia," *Europe-Asia Studies* 62, no. 7 (2010): 1177–1178; and Vjeran Pavlaković, "Yugoslav Volunteers in the Spanish Civil War," *Research Paper Series of Rosa Luxemburg Stiftung Southeast Europe* 4 (2016), esp. 35–38, 70.

53. For a recent addition to this debate, see Kirschenbaum, *International Communism*, 117–120, 137–144.

54. See, for instance, Gurney, *Crusade in Spain*, 144; and Jackson, *Fallen Sparrows*, 107–108. In variance with Gurney, Peter Carroll argues that executions were, in fact, very rare. Peter N. Carroll, "The Lincoln Brigade," *New York Review of Books*, 2 Mach 1995, http://www.nybooks.com/articles/1995/03/02/the-lincoln -brigade/.

55. Tom Buchanan, "Edge of Darkness: British 'Front-Line' Diplomacy in the Spanish Civil War, 1936–1937," *Contemporary European History* 12, no. 3 (2003): 289, 292; and Cecil D. Eby, *Comrades and Commissars: The Lincoln Brigade in the Spanish Civil War* (University Park, PA, 2006), 66, 82, 201, 383.

56. Orwell, *Homage to Catalonia*, 193.

57. Gurney, *Crusade in Spain*, 144, 152.

58. Gurney, *Crusade in Spain*, 161.

59. Kemp, *The Thorns of Memory*, 9.

60. Christopher R. Browning, *Ordinary Men: Reserve Police Battalion 101 and the Final Solution in Poland* (London, 2001), esp. 160, 184–186.

61. Gutmann, *Building a Nazi Europe*, 171; and Steffen Werther, review of Dennis Larsen and Therkel Stræde, *En skole i vold: Bobruisk 1941–1944. Frikorps Danmark og det tyske besættelsesherredømme i Hviderusland* (Copenhagen, 2014), in *Holocaust and Genocide Studies* 29, no. 3 (2015): 492–495.

62. Gutmann, *Building a Nazi Europe*, 57 and 170.

63. George H. Stein, *The Waffen SS: Hitler's Elite Guard at War* (Ithaca, NY, 1966), 154.

64. Stein, *The Waffen SS*, 155, 158–159.

65. Special Branch, "Statement of Witness," signed by Wallace Levey, 6 June 1948, TNA, HO 45 / 25587.

66. Special Branch, "Statement of Witness," signed by Stanley Jackson, 3 June 1948, TNA, HO 45 / 25587.

67. Ehud Yonay, *No Margin for Error: The Making of the Israeli Air Force* (New York, 1993), 55–56; and Joe Woolf, "Leonard Fitchett," World Machal website, http://

machal.org.il/index.php?option=com_content&view=article&id=178&Itemid=544&lang=en.

68. James Bruce, "Arab Veterans of the Afghan War," *Jane's Intelligence Review* 7, no. 4 (1995): 175.

69. Cited in Fawaz A. Gerges, "The End of the Islamist Insurgency in Egypt? Costs and Prospects," *Middle East Journal* 54, no. 4 (2000): 597. See also Magnus Ranstorp, "Interpreting the Broader Context and Meaning of Bin Laden's Fatwa," *Studies in Conflict & Terrorism* 21, no. 4 (1998): 321–330.

70. Blair, "Iraq, Syria and the Middle East."

71. Hegghammer, "Should I Stay or Should I Go?," 1–15.

72. Byman and Shapiro, "Be Afraid. Be a Little Afraid."

73. Jytte Klausen, "They're Coming: Measuring the Threat from Returning Jihadists," *Foreign Affairs*, 1 October 2014, https://www.foreignaffairs.com/articles/iraq/2014-10-01/theyre-coming.

74. Steven Mufson and William Booth, "How Two Brothers Became Suspected Members of the Paris Terror Plot," *Washington Post*, 17 November 2015, https://www.washingtonpost.com/world/europe/how-two-brothers-became-suspected-members-of-the-paris-terror-plot/2015/11/17/9551e9ce-8cc9-11e5-934c-a369c80822c2_story.html; and Julian Borger, Paul Scruton, Cath Levett, Paul Torpey, and Simon Jeffrey, "The Men Who Attacked Paris: Profile of a Terror Cell," *Guardian*, 18 March 2016, http://www.theguardian.com/world/ng-interactive/2015/nov/16/men-who-attacked-paris-profile-terror-cell.

75. Andrew Higgins and Maïa de la Baume, "Two Brothers Suspected in Killings Were Known to French Intelligence Services," *New York Times*, 8 January 2015, https://www.nytimes.com/2015/01/08/world/two-brothers-suspected-in-killings-were-known-to-french-intelligence-services.html?_r=0; Mohammed Ghobari, "Exclusive: Paris Attack Suspect met Prominent al Qaeda Preacher in Yemen—Intelligence Source," Reuters, 9 January 2015, http://www.reuters.com/article/us-france-shooting-yemen-idUSKBN0KI0PW20150109.

76. See, for instance, Carolin Goerzig and Khaled Al-Hashimi, *Radicalization in Western Europe: Integration, Public Discourse, and Loss of Identity among Muslim Communities* (Abingdon, UK, 2015), 2.

77. Peter R. Neumann, *Radicalized: The New Jihadists and the Threat to the West* (London, 2016), 92.

78. Richard Watson, "Briton 'Doing His Duty' by Fighting for Group Linked to al-Qaeda in Syria," BBC *Newsnight*, 21 November 2013, http://www.bbc.co.uk/news/uk-25022097.

79. Alexandra Sifferlin, "What the Paris Attacks 'Mastermind' Once Told an ISIS Magazine," *Time*, 16 November 2015, http://time.com/4114189/mastermind-behind-paris-attacks-previously-interviewed-by-isis-magazine/.

80. Paul Pickering, "A New Terror to Death: The Disappearance of John Dunmore Lang and Giuseppe Garibaldi," *Journal of Australian Studies* (forthcoming). See also "Scott, Andrew George (1842–1880)," *Australian Dictionary of Biography*, Australian National University website, http://adb.anu.edu.au/biography/scott-andrew-george-4546/text7451.

81. Cited in Yossi Melman and Eitan Haber, *The Spies: Israel's Counter-Espionage Wars* (Tel Aviv, 2002), 117 [Hebrew]. See also "Yisrael Bar (1961)," Israeli Security Agency website, http://www.shabak.gov.il/english/history/affairs/pages/yisraelbar.aspx. Bar was convicted of espionage and died in 1966 while serving his prison sentence.

8. Links in a Chain

1. Jackson, *Fallen Sparrows*, 5.

2. See, for instance, Carroll, *The Odyssey of the Abraham Lincoln Brigade*, esp. 209–368; Josie McLellan, *Antifascism and Memory in East Germany: Remembering the International Brigades 1945–1989* (Oxford, 2004). In Czechoslovakia and Hungary a number of prominent veterans of the Spanish Civil War were put on show trials and executed in the late 1940s and early 1950s. They fell victim to the Stalinist-inspired purges that engulfed those countries in the early Cold War. Rémi Skoutelsky, *Noveded en el frente: Las Brigadas Internacionales en la Guerra Civil* (Madrid, 2006), 439–442.

3. Varley, "Contesting Concepts of the Nation in Arms"; Nir Arielli and Davide Rodogno, "Transnational Encounters: Hosting and Remembering Twentieth-Century Foreign War Volunteers," *Journal of Modern European History* 14, no. 3 (2016): 315–320; and Jorge Marco and Peter Anderson, "Legitimacy by Proxy: Searching for a Usable Past through the International Brigades in Spain's Post-Franco Democracy, 1975–2015," *Journal of Modern European History* 14, no. 3 (2016): 391–410.

4. See Coulombe, *The Pope's Legion*, esp. 50; O'Duffy, *Crusade in Spain;* Gurney, *Crusade in Spain*.

5. Rubenfeld to Jewish Agency in Jerusalem, 8 December 1947, CZA, S25 / 8172. It is unclear which unit of the RAF Rubenfeld joined. See "Milton Rubenfeld," 101 Squadron website, http://101squadron.com/101real/people/rubenfeld.html. For more on Rubenfeld's service in the Israeli air force, see Ezer Weizman, *On Eagles' Wings: The Personal Story of the Leading Commander of the Israeli Air Force* (New York, 1976), 59–63.

6. Eric Hobsbawm, "Inventing Traditions: Introduction," in *The Invention of Tradition,* ed. Eric Hobsbawm and Terence Ranger (Cambridge, 1983), 1–14.

7. Edmundo Murray, "John Devereux (1778–1854), Army Officer and Recruiter for the Irish Legion in Simón Bolívar's Army," *Irish Migration Studies in Latin America* 4, no. 2 (2006): 93–94; Brown, *Adventuring through Spanish Colonies*, 215, 219n2; and Auguste Levasseur, *Lafayette in America in 1824 and 1825: Or, Journal of a Voyage to the United States*, vol. 2 (Philadelphia, 1829), 247.

8. Repousis, *Greek–American Relations*, 42–46.

9. Andrews Norton, *A Review of the Character and Writing of Lord Byron* (London, 1826), 137–138.

10. Cited in Doyle, *The Cause of All Nations*, 19.

11. Cited in Hall and Nordhoff, *The Lafayette Flying Corps*, vol. 1, 6.

12. Flammer, *The Vivid Air*, x, 182.

13. "The Eagle Takes the Air," *Flight*, 17 October 1940, 331.

14. Cited in Lynne Olson and Stanley Cloud, *For Your Freedom and Ours: The Kościuszko Squadron: Forgotten Heroes of World War II* (London, 2003), 29.

15. Murray, *Wings over Poland*, vii–viii.

16. Lord Byron, *The Age of Bronze; or, Carmen Seculare et Annus Haud Mirabilis* (London, 1823), 12.

17. St. Clair, *That Greece Might Still Be Free*, 337; and Repousis, *Greek–American Relations*, 44–45. See also Samuel G. Howe, *An Historical Sketch of the Greek Revolution* (New York, 1828).

18. Eugene Michail, *The British and the Balkans: Forming Images of Foreign Lands, 1900–1950* (London, 2011), 58.

19. Hopkins, *Into the Heart of Fire*, 105–106. See also Wintringham, *English Captain*, 17.

20. Baxell, *Unlikely Warriors*, 116.

21. *British Battalion XV International Brigade: Memorial Souvenir* (London, [1939]).

22. Karma Nabulsi, *Traditions of War: Occupation, Resistance and the Law* (Oxford, 2005), 223.

23. Alex Storozynski, *The Peasant Prince: Thaddeus Kosciuszko and the Age of Revolution* (New York, 2009), 283.

24. Cited in Koropeckyj, *Adam Mickiewicz*, 384.

25. Koropeckyj, *Adam Mickiewicz*, 398–407.

26. Adam Zamoyski, *The Polish Way: A Thousand-Year History of the Poles and Their Culture* (London, 1987), 277, 288; Jonathan Sperber, *The European Revolutions, 1848–1851* (Cambridge, 1994), 224; and Norman Davies, *God's Playground: A History of Poland*, vol. 2 (Oxford, 2005), 26, 252.

27. Helen Graham, *The Spanish Civil War: A Very Short Introduction* (Oxford, 2005), 44.

28. Zamoyski, *The Polish Way*, 359; Olson and Cloud, *For Your Freedom*, 6, 74, 89, 91, 151.

29. President Clinton's speech at Warsaw's Castle Square, 10 July 1997, C-Span, http://www.c-span.org/video/?87640-1/nato-expansion.

30. Cited in Murphy, "The Wild Geese," 24.

31. Cited in Brown, *Adventuring through Spanish Colonies,* 116.

32. Michael G. Connaughton, "Beneath the Emerald Green Flag: The Story of Irish Soldiers in Mexico," *Society for Irish Latin American Studies,* September 2005, http://www.irlandeses.org/sanpatriciosA.htm; Bartlett and Jeffries, "An Irish Military Tradition?," 20; Coulombe, *The Pope's Legion,* 55; and Macnab, *The French Colonel,* 165–166.

33. Hubert Gough, "An Irish Brigade," *Times* [London], 26 September 1941, 5.

34. John M. Andrews to Churchill, 12 December 1941, TNA, PREM 3 / 129 / 5.

35. Bartlett and Jeffries, "An Irish Military Tradition?," 20; O'Connor, *Irish Officers in the British Forces,* 116–117.

36. Callivretakis, "Les Garibaldiens," 163–177; Pécout, "Philhellenism in Italy," 413; and Riall, *Garibaldi,* 347–350.

37. Riall, *Garibaldi,* 355.

38. Basso (Chambéry) to Visconti Venosta, 8 October 1870, *Documenti Diplomatici Italiani (DDI),* series 2, vol. 1, 178; and De Launay (Berlin) to Visconti Venosta, 15 February 1871, *DDI,* series 2, vol. 2, 179–181.

39. Terzuolo, "The Garibaldini in the Balkans," 115–117.

40. Callivretakis, "Les Garibaldiens," 163; and Pécout, "Philhellenism in Italy," 419.

41. Richard C. Hall, *The Balkan Wars 1912–1913: Prelude to the First World War* (London, 2000), 63.

42. Giuseppe Garibaldi, *A Toast to Rebellion* (London, 1936), 2.

43. Cited in "Peppino Garibaldi Plans Fight for Independence," *San Francisco Call,* 10 July 1911, 10.

44. Garibaldi, *A Toast to Rebellion,* 31.

45. Garibaldi, *A Toast to Rebellion,* 41.

46. Garibaldi, *A Toast to Rebellion,* 51.

47. Macnab, *The French Colonel,* 163–164; "The Man Who Captured Churchill," *The Italic Way* 26 (1996): 20.

48. Zeffiro Ciuffoletti, Arturo Colombo, and Annita Garibaldi Jallet, *I Garibaldi dopo Garibaldi: La tradizione familiare e l'eredità politica* (Manduria, Italy, 2005), 155.

49. Garibaldi, *A Toast to Rebellion,* 73.

50. Garibaldi, *A Toast to Rebellion,* 172.

51. Lawrence D. Taylor, "The Great Adventure: Mercenaries in the Mexican Revolution, 1910–1911," *The Americas* 43, no. 1 (1986): 25–45.

52. Heyriès, "The Garibaldian Volunteers," 371; and Acciai, *Antifascismo volontario,* 152.

53. Pietro Borghi, "Garibaldi visto da un garibaldino," *Il Garibaldino*, no. 5 (27 July 1937): 3, Russian State Archive of Social and Political History, Comintern Archives, F. 545, Op. 3, D. 193.

54. "Note," 15 October 1939, Archives Nationales de France (ANF), folder F / 7 / 14748, subfolder "Unione Popolare Italiana"; Robert Gerwarth and Lucy Riall, "Fathers of the Nation? Bismarck, Garibaldi and the Cult of Memory in Germany and Italy," *European History Quarterly* 39, no. 3 (2009): 400; and Hubert Heyriès, *Les Garibaldiens de 14: Splendeurs et misères des chemises rouges en France de la Grand Guerre à la Seconde Guerre mondiale* (Nice, 2005), 383–396.

55. Riall, *Garibaldi*, 6; Philip Morgan, *The Fall of Mussolini: Italy, the Italians and Second World War* (Oxford, 2008), 178–181, 208; and Stefano Gestro, *La divisione Italian partigana Garibaldi* (Milan, 1981), esp. 347–353.

56. Steiner, *Die Freiwilligen*, 16–19. The Nazis, it should be noted, held Sante Garibaldi at the Dachau concentration camp during the Second World War.

57. "La nostra storia," Associazione Nazionale Veterani e Reduci Garibaldini website, http://nuke.garibaldini.com/ANVRG/Lanostrastoria/tabid/469/Default .aspx. The association does not include the communist partisan Garibaldi formations that fought in Italy during the Second World War within its remit.

58. Tony Barber, "Croatian Serbs 'Recruit Italian Fighters': 'Garibaldi Unit' Defies Rome and Attempts to Take Back Land Lost to Yugoslavia after Second World War," *Independent*, 20 October 1993, http://www.independent.co.uk/news/world /europe/croatian-serbs-recruit-italian-fighters-garibaldi-unit-defies-rome -and-attempts-to-take-back-land-1512064.html.

59. Walter Citrine, *My Finnish Diary* (Harmondsworth, UK, 1940), 162.

60. Susan D. Pennybacker, *From Scottsboro to Munich: Race and Political Culture in 1930s Britain* (Princeton, NJ, 2009), 238.

61. Paul Eduard Koch to the editor of *Picture Post*, 30 January 1940, ANF, Willi Münzenberg collection, folder F 7 / 15125, subfolder no. 3.

62. Cited in Bercuson, *The Secret Army*, 51.

63. Joe Woolf, "Jules Cuburnek—Air Force Navigator," World Machal website, http://www.machal.org.il/index.php?option=com_content&view=article&id =208&Itemid=294&lang=en.

64. Levett, *Flying under Two Flags*, 127.

65. Author's interview with Shmuel Segal, Tel Aviv, 7 July 2009; and Eran Torbiner, "In Memory of Shmuel Segal (1917–2012)," *Hagada Hasmalit*, 18 January 2012, http:// hagada.org.il/2012/01/18/%D7%9C%D7%96%D7%9B%D7%A8%D7%95 -%D7%A9%D7%9C -%D7%A9%D7%9E%D7%95%D7%90%D7%9C -%D7%A1%D7%92%D7%9C-1917-2012/[Hebrew].

66. For instance, in October 1968 Evgenyi Tyazhel'nikov, the First Secretary of the Soviet Union's youth organization, the Komsomol, delivered a speech on the

organization's contribution to struggles in the developing world. He spoke of a revolutionary thread that linked the 1930s, the defeat of the Nazis in the 1940s and volunteers who headed out to Africa, Asia, and Latin America in the 1950s and beyond: "Fulfilling their international duty, Soviet youth volunteers joined the battle against fascism, for the freedom of republican Spain." Robert Hornsby, "The Post-Stalin Komsomol and the Soviet Fight for Third World Youth," *Cold War History* 16, no. 1 (2016): 83.

67. Bill Bailey, Foreword to Jones, *Brigadista,* xvi; emphasis in the original.

68. Bailey, Foreword, xvii.

69. Author's interview with Gaston Besson, Vinkovci, Croatia, 14 March 2010. For more on how Rózsa-Flores likened his men to the International Brigades and quoted from a speech by the Spanish Communist leader Dolores Ibárruri, see the film *Chico,* dir. Ibolya Fekete (2001).

70. See, for instance, the comments made by the volunteer "Jean-Philippe" from France, who was interviewed by a French news crew during the war: "Internacionalna 108 HVO brigade—Doku film HR," https://wn.com/photos_album_108_hvo_part_1.

71. Cited in Marc Sageman, *Leaderless Jihad: Terror Networks in the Twenty-First Century* (Philadelphia, 2008), 8. See also Sean O'Neill, "Britons Fall Victim to an Islamic Dream," *Telegraph,* 5 October 2001, http://www.telegraph.co.uk/news/uknews/1358541/Britons-fall-victim-to-an-Islamic-dream.html.

72. "Igor Girkin, krivac za rušenje malezijskog aviona, ratovao i u BiH," Klix.ba, 25 July 2014, http://www.klix.ba/vijesti/svijet/igor-girkin-krivac-za-rusenje-malezijskog-aviona-borio-se-i-u-ratu-u-bih/140718101; and Gianluca Mezzofiore, "Igor Strelkov's Bosnian Diary: Ukraine Separatist Leader's 1992 Bloody War," *International Business Times,* 29 July 2014, http://www.ibtimes.co.uk/igor-strelkovs-bosnian-diary-ukraine-separatist-leaders-1992-bloody-war-1458897.

73. See, for instance, "Sa Krima na TV Happy, preko Koteža," 26 March 2014, http://www.e-novine.com/srbija/100986-Krima-Happy-preko-Kotea.html.

74. Umberto Bacci, "Ukraine Crimea Crisis: Serb Chetnik Militia Joins Pro-Russian Patrols," *International Business Times,* 10 March 2014, http://www.ibtimes.co.uk/ukraine-crimea-crisis-serb-chetnik-militia-joins-pro-russian-patrols-1439654.

75. "Bivši glasnogovornik srpskog MUP-a otišao se boriti u Ukrajinu," Klix.ba, 16 September 2014, http://www.klix.ba/vijesti/regija/bivsi-glasnogovornik-srpskog-mup-a-otisao-se-boriti-u-ukrajinu/140916069.

76. Leo Marić, "Hrvatski dragovoljci idu braniti Ukrajinu, evo što poručuju Hrvatima," *Sloboda,* 6 February 2015, http://www.sloboda.hr/hrvati-u-ukrajini/.

77. "Hrvatski dobrovoljci u Ukrajini: Došli smo da ubijamo četnike iz Srbije," Klix.ba, 4 April 2015, http://www.klix.ba/vijesti/svijet/hrvatski-dobrovoljci-u-ukrajini-dosli-smo-da-ubijamo-cetnike-iz-srbije/150404089?utm_source=facebook.

78. Hegghammer, "The Rise of Muslim Foreign Fighters," 72; Hamid and Farrall, *The Arabs at War in Afghanistan,* 183–184; and Bakke, "Help Wanted?," 166–167, 169.

79. Li, "'Afghan Arabs,' Real and Imagined."

80. Cited in Aryn Baker, "The Abdullah Azzam Brigades: Behind the Terrorist Group That Bombed Iran's Beirut Embassy," *Time,* 20 November 2013, http://world.time.com/2013/11/20/the-abdullah-azzam-brigades-behind-the-group-that-bombed-irans-beirut-embassy/. For more on the Abdullah Azzam Brigades, see Rohan Gunaratna and Aviv Oreg, *The Global Jihadi Movement* (Lanham, MD, 2015), 239–241; and "Lebanon 'Arrests Head of Abdullah Azzam Brigades,'" BBC News, 1 January 2014, http://www.bbc.co.uk/news/world-middle-east-25566784.

Epilogue

1. See Prelinger, "Less Lucky than Lafayette," 266; Skoutelsky, "L'engagement des volontaires," 11; Francis, "Welsh Miners and the Spanish Civil War," 191; Petrou, *Renegades,* 13; Hutt, *Paint,* 108–110; Felter and Fishman, *Al-Q'aida's Foreign Fighters in Iraq,* 16; and Percy, "Meet the American Vigilantes."

2. "Why an Ordinary Man Went to Fight Islamic State," *Economist,* 24 December 2016, http://www.economist.com/news/christmas-specials/21712055-when-islamic-state-looked-unbeatable-ordinary-men-and-women-went-fight-them-why.

3. For the political theorist and philosopher Hannah Arendt, who delves into the classical concept of *vita activa* (active life), "to act, in its most general sense, means to take an initiative, to begin . . . to set something in motion." Hannah Arendt, *The Human Condition* (Chicago, 1998 [1958]), 177.

4. Percy, "Meet the American Vigilantes."

5. Matt Blake, "Briton Killed while Fighting against Isis in Syria," *Guardian,* 2 January 2017, https://www.theguardian.com/uk-news/2017/jan/02/briton-ryan-lock-killed-while-fighting-against-isis-in-syria.

6. Jim Garamone, "Defeat-ISIS 'Annihilation' Campaign Accelerating, Mattis Says," US Department of Defense, 28 May 2017, https://www.defense.gov/News/Article/Article/1196114/defeat-isis-annihilation-campaign-accelerating-mattis-says/.

Acknowledgments

Historians tend to specialize in specific periods and regions. My own research thus far has focused on fairly bounded episodes in the history of Europe, the Middle East, and North Africa between the 1920s and the 1990s. Therefore, when I set out to write the history of a global phenomenon that spans nearly two and a half centuries and is still very relevant today, I turned to a good number of people for advice. I consider myself fortunate to work at the School of History at the University of Leeds, where many of my colleagues were willing to make sensible suggestions, provide me with useful historical sources, and comment on various parts of my manuscript. I am particularly grateful to Enrico Acciai, Holger Afflerbach, Stephen Alford, Peter Anderson, Simon Ball, Manuel Barcia Paz, Adam Cathcart, Kate Dossett, Claire Eldridge, Robert Hornsby, Elisabeth Leake, and Jessica Meyer. I would also like to thank William Gould, the school's former director of research, for his part in making sure I had the time and the means needed to get the research for this book done.

I also consider myself fortunate to have good students. Postgraduate student Kristo Karvinen was of great help with all things Finnish. I am very grateful to all the students, past and present, who took my third-year course HIST3743, which examines historical and contemporary transnational war volunteers. They were (and are still) a constant source for ideas and new ways of looking at the phenomenon. A special thanks goes to my former students Hannah Graham, Katie Milne, and Rebecca Stead for helping me to carry out the survey that is summarized in the Epilogue.

I am similarly grateful to friends and colleagues who work in several other institutions and who have contributed to this research project in various important ways. These include Marco Bresciani, Čarna Brković, Garry Campion, Francesco Capello, Gabriela Frei, Martin R. Gutmann, Gilad Heller, Hubert Heyriès, Graciela Iglesias Rogers, Bojan Kovačić, Stefano Manzo, OSCE's John Martin, David Motadel, Steven O'Connor, Derek Penslar, Guy Perry, Niccolò Petrelli, Paul Pickering, Fraser Raeburn, Shiran Reichenberg, Davide Rodogno, Arthur Scott-Geddes, Ewelina Siemianowska, Marie-Cecile Thoral, Inge Van Hulle, and Louis Vine.

I am grateful to those former foreign volunteers who agreed to share their experiences with me. I would especially like to thank Roland Bartetzko, Ivan Farina, and Antaine Mac Cosciar for their enduring patience. My heartfelt thanks goes out to Valerie Carder, who was willing to share her thoughts with me not long after her son, John Gallagher, was killed while fighting alongside Kurdish forces in Syria.

I would like to express my gratitude to my editor at Harvard University Press, Ian Malcolm, for believing in the potential of this project when others didn't, and for all his support throughout the publishing process. I am also grateful to the anonymous referees for their constructive comments.

Finally, I would like to thank my family: my parents for their constant support and interest in my work, my two wonderful children for bringing so much light into my life (especially when they wake me up at five o'clock in the morning), and my Vanja, for everything. In fact, the idea to research foreign war volunteers first emerged from a conversation we had, back in Jerusalem, almost fifteen years ago.

Index